QoS in Wireless Sensor/Actuator Networks and Systems

QoS in Wireless Sensor/Actuator Networks and Systems

Special Issue Editor

Mário Alves

MDPI • Basel • Beijing • Wuhan • Barcelona • Belgrade

MDPI

Special Issue Editor
Mário Alves
School of Engineering (ISEP/IPP)
Portugal

Editorial Office
MDPI
St. Alban-Anlage 66
4052 Basel, Switzerland

This is a reprint of articles from the Special Issue published online in the open access journal *Journal of Sensor and Actuator Networks* (ISSN 2224-2708) from 2017 to 2018 (available at: https://www.mdpi.com/journal/jsan/special_issues/QoS_netw_syst)

For citation purposes, cite each article independently as indicated on the article page online and as indicated below:

LastName, A.A.; LastName, B.B.; LastName, C.C. Article Title. *Journal Name* **Year**, *Article Number, Page Range.*

ISBN 978-3-03897-362-1 (Pbk)
ISBN 978-3-03897-363-8 (PDF)

Contents

About the Special Issue Editor

Mário Alves was born in 1968 and earned a PhD (2003) in Electrical and Computer Engineering from the University of Porto, Portugal. He is a Professor in ECE at Politécnico do Porto (ISEP/IPP). He worked as a Research Associate at the CISTER Research Unit from 1997 to 2015. At the beginning of his career, he participated in several international projects related to industrial communication systems, developing his MSc and PhD in synergy with a couple of these projects (CNMA, RFieldbus). He has served as a reviewer and published in top-notch conference proceedings (e.g. RTSS, ECRTS, ICDCS, SECON, MASS, EWSN) and journals (e.g. ACM TOSN, IEEE TII, Elsevier ComNet/Automatica/Ad-Hoc Net, Springer RTSJ) in his areas of expertise areas. He also has received best paper awards (e.g. ECRTS'07) and supervised the EWSN'09 best MSc Thesis Award. He has actively participated in the organization of many international conferences and workshops (e.g. RTSS, EWSN, ECRTS, IEEE WFCS, CONET) and projects (namely, lead WP4 in the EMMON ARTEMIS project, the COTS4QoS research cluster in the CONET NoE, the open-ZB framework, and the MASQOTS FCT project). His personal research interests are mainly devoted to supporting quality-of-service (QoS) in large-scale and dense WSNs, and to innovative teaching/self-learning tools/methodologies for electrical circuit analysis.

Journal of
*Sensor and
Actuator Networks*

MDPI

Editorial

Special Issue: Quality of Service in Wireless Sensor/Actuator Networks and Systems

Mário Alves

Politécnico do Porto, School of Engineering (ISEP/IPP), Rua Dr. António Bernardino de Almeida, 431, 4249-015 Porto, Portugal; mjf@isep.ipp.pt; Tel.: 22-834-0500

Received: 23 February 2018; Accepted: 23 February 2018; Published: 25 February 2018

1. Introduction

Wireless communications have long been a part of our daily lives: from remote controls to baby monitors, from cellular to local area networks, connecting us to the Internet and integrating our various gadgets and peripherals. However, industrialists have been very conservative in allowing wireless to enter their factory automation and process control systems. This has been mainly because "wireless" technologies had only been used in very specific and limited industrial scenarios (e.g., to facilitate point-to-point connections) and often granted insufficient reliability, security, or lifetime guarantees.

As wireless technology matured (e.g., IEEE 802.11-based and IEEE 802.15.4-based), wireless sensors, actuators, and controllers started penetrating the factory floor, complementing or even replacing their wired counterparts. While these "hybrid" wired/wireless systems are already a commodity, we are at the dawn of a new era, where computers, sensors, and actuators are becoming increasingly embedded and ubiquitous, scaling up systems at unprecedented levels. The number of (embedded, networked) devices will grow dramatically, while the size of individual devices/nodes will necessarily shrink. The so-called "Internet-of-Things" and "Industry 4.0" are very broad umbrellas that fit into this paradigm.

Wireless sensor/actuators networks (WSANs) are thus being increasingly used in a panoply of applications, such as industrial automation, process control, ambient assisted living, structural health monitoring, and homeland security. Most of these applications require specific quality-of-service (QoS) guarantees from their underlying communication infrastructures (regardless of their wireless, wired, or hybrid nature).

While QoS has been traditionally associated with bit/data rate, network throughput, message delay, and bit/packet error rate, these properties alone do not reflect the overall "quality of service" that needs to be provided for such applications (and for their users). Other (non-functional) properties such as scalability, security, mobility, or energy sustainability must also be considered in the design of such complex cyber-physical systems. Importantly, these properties often conflict.

This special issue targeted contributions in wireless sensor/actuator networks and systems (WSANSs) addressing QoS properties (hopefully in combination) such as reliability and robustness, timeliness and real-time, scalability, mobility, security and privacy, and energy efficiency and sustainability. We sought for works that were sufficiently mature, tested, and evaluated through analytical, simulation, or experimental models. Extensions to previously published works were eligible, provided that this fact was clearly stated in the submission and that the new contribution was significant.

We were envisaging works covering one or more of the following WSAN topics, with QoS as an overarching aspect:

-    System architectures: e.g., improving hardware (e.g., radio technology), software (including operating systems), and communication network architectures to achieve better QoS; scalability; WSAN integration in, and interoperability with, legacy wired systems; cross-layer design.

- Reliability and robustness: improving communication errors detection/correction, hardware robustness, and systems reliability in general.
- Timeliness and real-time: improving the timing behavior and reducing/bounding (end-to-end) communication delays; innovative time synchronization techniques.
- Security and privacy: new mechanisms to grant adequate levels of security/privacy without jeopardizing energy and time.
- Mobility: mechanisms to support mobile devices in a seamless and transparent way, while still respecting overall QoS requirements.
- Energy sustainability, efficiency, and harvesting: improving devices/system lifetime, e.g., through optimized communications scheduling/duty-cycling and energy/delay trade-offs.
- Radio interference identification and mitigation: improving the detection, classification, and mitigation of communication errors deriving from radio propagation and interference.
- Communication and network protocols: QoS add-ons; performance/worst-case analysis (analytical, simulation, and experimental).
- QoS in the Internet-of-Things, Cyber-Physical Systems, and Industry 4.0 contexts.
- Experimental facilities and test-beds, pilot demonstrations/deployments; innovative simulation and emulation models, platforms, and methodologies.
- Real-world applications, such as in smart health, environmental/structural monitoring, factory automation, process control, smart buildings, body sensor networks, vehicular networks, or security/surveillance.
- Communication standards and technologies for WSAN and LPWAN, e.g., IEEE 802.11, WiFi, IEEE 802.15.4, ZigBee, 6loWPAN, WirelessHART, ISA SP100, LoRa, SigFox, and Narrowband-IoT and their integration/interoperability with wired networks and with the Internet/Cloud.
- Novel communication technologies to overcome an increasingly overcrowded radio spectrum (e.g., visible light, mm-wave, thermal, vibration, acoustic) communication.

2. Summary of Contributions

This special issue gathers together an extremely rich set of contributions, addressing several WSAN domains and sharing QoS as a common denominator. Eight papers have made it through a rigorous and iterative peer-reviewing process (three reviews per paper, at least two review rounds), involving 38 authors from all over the world (North and South America, Europe, Asia, and Australia) from academia, industry, and the military. Each paper features at least one reference author which is highly reputed in this scientific domain, totaling over one hundred thousand citations altogether.

Papers cover a wide range of topics, namely the optimization of retransmission scheduling in IEEE 802.15.4e WSANs [1], an experimental evaluation of LoRa reliability [2], the estimation of WSAN lifetime based on innovative battery models [3], a novel radio interference classification method for WSANs [4], a dynamic QoS-aware MAC that can be boosted for long-range communications [5], an RSSI-based model-learning for target localization/tracking [6], using sensor network calculus for designing WSANs with predictable e2e delays [7], and decision-centric WSAN resource management [8]. A brief summary of each paper is provided here.

The first paper to be published in this special issue, [1], investigates how the reliability of Industrial WSANs based on the IEEE 802.15.4e LLDN (low-latency deterministic network) protocol can be increased. The authors explore the inherent characteristics of the protocol to reschedule predetermined retransmission slots in a clever way, improving the reliability of uplink (sensor nodes -> coordinator) traffic and maximizing the probability of correct packet delivery. Importantly, two basic retransmission scheduling mechanisms are proposed that are backward compatible with the standard protocol, while two other approaches relying on cooperative relayers lead to more substantial gains but require minor changes to the standard. The authors build on two channel models: one static (fixed randomly chosen

PER) and another one time-varying (hopping between two randomly chosen PERs) and evaluate the performance (essentially the superframe packet success rate) using a custom-made simulator.

The authors of [2] report the main results of an extensive experimental evaluation of the reliability of LoRa, one of the most prominent long-range low-power WSAN to date, along with SigFox and Narrowband-IoT. The authors focus on the impact of physical layer settings and ambient temperature on effective data rate and energy efficiency. The authors build on indoor, outdoor, and underground experiments, concluding, e.g., that, when sensor nodes are at the boundaries of their communication range, it is better to use higher bit rates together with a retransmission scheme, rather than selecting a slower setting that maximizes PDR. It is also systematically shown that there is a direct correlation between ambient temperature, PDR, and received radio signal strength, for different LoRa radios. Over a range of 60 °C, the received signal strength consistently decreases 1 dBm/10 °C, which has a great impact for nodes at the edge of the communication range: a perfectly good link at 15 °C becomes unusable at 60 °C.

WSANs are typically dependent (at least partially) on battery-powered devices. Therefore, estimating their current energy capacity and their lifetime may be of paramount importance, namely for energy-aware protocols and resource management. However, operating temperature and discharge current variations have a great impact on most electrochemical batteries' capacity, rendering the estimation of their voltage/charge behavior over time a tricky task. The authors of [3] propose an analytical model to estimate the state-of-charge and voltage of batteries based on their temperature, avoiding more expensive and energy-consuming hardware-based solutions that may compromise WSANs scalability. The authors implement, evaluate, and validate their analytical model through experiments over six different low-power low-memory COTS WSAN nodes (from the Atmel ATmega and SAM families).

Most WSAN technologies operate in the license-free ISM band, so communications are prone to external and cross-technology interference, namely from other wireless networks (e.g., WiFi and Bluetooth) and a myriad of daily use devices sharing the same spectrum (e.g., microwave ovens, baby monitors, cordless phones, audio/video transmitters, and remote controls). While some WSAN applications may cope with lost packets and extra communication delays, when we look into industrial WSAN application contexts, things may radically change, as reliability and timeliness are usually at stake. In this context, the authors of [4] describe a novel method for classifying interference sources in IEEE 802.15.4-based IWSANs, which may then be complemented with interference mitigation techniques. This scheme builds on a machine learning technique (support vector machines) for classifying interference from IEEE 802.11 networks and microwave ovens, as well as the presence of interference-free channels. Extensive tests in three industrial scenarios show that a high classification accuracy (over 80%) can be obtained in a very short channel sensing time (below 300 ms). The computational effort and memory footprint are very small, enabling this classifier to be implemented in low-cost COTS WSAN hardware.

Unmanned aerial vehicles (UAVs) are becoming a commodity. While up to a decade ago they were used for very specific applications such as search & rescue and homeland security), nowadays it is increasingly common to use such devices for environmental monitoring, forest surveillance, 3D-mapping, and companies/sites promotion. Some scenarios may involve several UAVs intercommunicating and interoperating among each other (in swarms) and with fixed and mobile infrastructures, devices, and people on the ground. In such multi-agent cyber-physical ecosystems, localization/tracking and wireless communications are paramount and extremely challenging, considering that QoS must be guaranteed in highly dynamic, uncontrolled, open-space, and large-scale settings. The authors of [5,6] engineer solutions toward this end.

The authors of [5] elaborate on a cooperative MAC protocol specifically designed for networks of UAVs, supporting traffic differentiation between best-effort (on-demand) and guaranteed bandwidth (periodic). The authors complement this data-link layer protocol with two mechanisms that improve the reliability and range of communications, namely an adaptive antenna array that enables omnidirectional and directional transmissions and a cooperative relayer mechanism. The proposed mechanism is evaluated through a probabilistic model (Markov chain) and simulations (ns2 and MatLab), proving its efficiency against four reference protocols.

RSSI-based object/target localization and tracking mechanisms usually feature a training phase for mitigating the particularities of each environment and optimizing their accuracy. The authors of [6] propose two run-time (on-line) model-learning mechanisms that overcome the cons of pre-run-time (off-line) alternatives, namely in terms of (wasted) time, flexibility, and adaptability to changes in highly dynamic scenarios. Both mechanisms have been implemented and experimentally evaluated in a 2D test-bed with a set of anchor nodes and mobile robots, showing significant improvements over traditional RSSI-range models.

A priori determination of WSAN performance is quite challenging, particularly when we aim at achieving the best energy-bandwidth tradeoff (worst-case dimensioning) in large-scale networks, considering the particularities of the communication stack (e.g., MAC and routing protocols non-determinism) and the dynamics of the environment (topology, obstacles, mobility, and EMI). It is fundamental to check if/how WSAN application requirements can be met, for instance concerning timeliness (e.g., maximum end-to-end delay of 3 s), reliability (a packet error rate below 1%), and scalability (a network must support up to 1000 nodes and cover an area with a 2 km radius), so that network parameters such as bit rate, nodes'/clusters' duty-cycle, TDMA slots reservation, and routers' buffer size can bed tune/optimized. The authors of [7] wrap up relevant work on *sensor network calculus*, a mathematical methodology that has been tailored (actually by the first author, over a decade ago) for worst-case analysis and dimensioning in sensor networks. The paper describes tools and applications of sensor network calculus and points out new research directions, such as encompassing stochastic models, downstream control traffic, and mobility support.

Last (but definitely not least), the authors of [8] address decision-centric resource management in QoS-aware WSAN systems, enabling the optimization of communication protocols, data storage, and scheduling policies for critical data collection, specifically for meeting decision needs. Toward this end, resource management heuristics/algorithms are proposed that meet at least the three following conditions: (i) collected data must have enough quality to support a user's decision; (ii) it should be recent (fresh) enough; (iii) the user's decision (based on the collected data) must meet the decision deadline. What sensors need to be activated to take the best decision at the lowest possible (resources) cost? How often do nodes need to sense the physical world and to communicate data? In which order (scheduling) should they be sampled? The proposed methodology and architecture (dubbed *Athena*) answers these questions, and has been evaluated (via simulation) considering a post-disaster route-finding scenario, proving to outperform traditional alternatives.

3. Conclusions

This special issue features eight extremely interesting and top-quality papers, browsing different scientific and technological issues related to WSANs, but all of them gravitate around a major concern: the Quality-of-Service (QoS) of communication networks and, in a broader perspective, of overall WSAN-based systems (hardware, software, and communication components).

It is widely accepted that the *non-functional* (QoS) properties of computing systems (namely WSAN systems) are at least as important as their *functional* counterparts. Guaranteeing that all QoS requirements are met is a challenging task, particularly considering their "conflicting" nature. For instance, guaranteeing that the end-to-end communication delay of a certain message stream is always smaller than a certain deadline may conflict with node/system lifetime requirements, since increasing the bit rate and/or the duty-cycle (of routing nodes) will demand more energy; increasing security may lead to extra energy consumption and processing/communication delays, due to the intrinsic algorithms and longer (encrypted) messages. All eight papers encompass these concerns, in one way or another.

J. Sens. Actuator Netw. **2018**, *7*, 9

We are moving into the realm of a new era, where ubiquitous computing is factually entering our daily lives, through different vests (e.g., "smart" phones, appliances, homes, cars, cities), but all share the need for underlying WSAN infrastructures. The scalability of these systems/applications, the close interaction with the physical world, and the dynamics of the agents involved and of the environment, will continue to challenge the exceptional researchers that have contributed to this special issue. I thank them here for their hard work and dedication and for their perseverance in revising and fine-tuning the papers toward their camera-ready versions.

Acknowledgments: I would like to thank the editor-in-chief Dharma Agrawal for his invitation to guest edit this special issue, particularly the flexibility to fine-tune the title and focus and to select invited authors/papers. I am also very grateful to all of the people in the editorial office I interacted with, particularly Rui Zuo and Louise Liu, as well as to all the anonymous reviewers, which have all contributed to the success of this project.

Conflicts of Interest: The authors declare no conflict of interest.

References

1. Willig, A.; Matusovsky, Y.; Kind, A. Relayer-Enabled Retransmission Scheduling in 802.15.4e LLDN—Exploring a Reinforcement Learning Approach. *J. Sens. Actuator Netw.* **2017**, *6*, 6. [CrossRef]
2. Cattani, M.; Boano, C.A.; Römer, K. An Experimental Evaluation of the Reliability of LoRa Long-Range Low-Power Wireless Communication. *J. Sens. Actuator Netw.* **2017**, *6*, 7. [CrossRef]
3. Rodrigues, L.M.; Montez, C.; Budke, G.; Vasques, F.; Portugal, P. Estimating the Lifetime of Wireless Sensor Network Nodes through the Use of Embedded Analytical Battery Models. *J. Sens. Actuator Netw.* **2017**, *6*, 8. [CrossRef]
4. Grimaldi, S.; Mahmood, A.; Gidlund, M. An SVM-Based Method for Classification of External Interference in Industrial Wireless Sensor and Actuator Networks. *J. Sens. Actuator Netw.* **2017**, *6*, 9. [CrossRef]
5. Gao, C.; Zeng, B.; Lu, J.; Zhao, G. Dynamic Cooperative MAC Protocol for Navigation Carrier Ad Hoc Networks: A DiffServ-Based Approach. *J. Sens. Actuator Netw.* **2017**, *6*, 14. [CrossRef]
6. Ramiro Martínez-de Dios, J.; Ollero, A.; Fernández, F.J.; Regoli, C. On-Line RSSI-Range Model Learning for Target Localization and Tracking. *J. Sens. Actuator Netw.* **2017**, *6*, 15. [CrossRef]
7. Schmitt, J.; Bondorf, S.; Poe, W.Y. The Sensor Network Calculus as Key to the Design of Wireless Sensor Networks with Predictable Performance. *J. Sens. Actuator Netw.* **2017**, *6*, 21. [CrossRef]
8. Lee, J.; Marcus, K.; Abdelzaher, T.; Amin, M.T.A.; Bar-Noy, A.; Dron, W.; Govindan, R.; Hobbs, R.; Hu, S.; Kim, J.-E.; et al. Athena: Towards Decision-Centric Anticipatory Sensor Information Delivery. *J. Sens. Actuator Netw.* **2018**, *7*, 5. [CrossRef]

Journal of
Sensor and
Actuator Networks

MDPI

Article

Relayer-Enabled Retransmission Scheduling in 802.15.4e LLDN—Exploring a Reinforcement Learning Approach

Andreas Willig *, Yakir Matusovsky and Adriel Kind

Department of Computer Science and Software Engineering/Wireless Research Centre,   ·
University of Canterbury, Private Bag 4800, Christchurch 8041, New Zealand; yakir.m@gmail.com (Y.M.);
adriel.kind@canterbury.ac.nz (A.K.)
* Correspondence: andreas.willig@canterbury.ac.nz; Tel.: +64-3-364-2987 (ext. 7869)

Academic Editor: Mário Alves
Received: 7 April 2017; Accepted: 31 May 2017; Published: 3 June 2017

Abstract: We consider the scheduling of retransmissions in the low-latency deterministic network (LLDN) extension to the IEEE 802.15.4 standard. We propose a number of retransmission schemes with varying degrees of required changes to the LLDN specification. In particular, we propose a retransmission scheme that uses cooperative relayers and where the best relayer for a source node is learned using a reinforcement-learning method. The method allows for adapting relayer selections in the face of time-varying channels. Our results show that the relayer-based methods achieve a much better reliability over the other methods, both over static (but unknown) and over time-varying channels.

Keywords: IEEE 802.15.4e; LLDN; retransmission scheduling; cooperative relayers; reinforcement learning

1. Introduction

Industrial wireless sensor networks (IWSN) have recently received considerable attention [1–3]. In particular, the IEEE 802.15.4 standard [4] is of great interest to researchers and practitioners, not the least due to its maturity and the commercial availability of components [5]. In 2012, the IEEE approved the IEEE 802.15.4e amendment to the IEEE 802.15.4 standard [6,7]. The amendment includes the *low latency deterministic network* (LLDN) extension, which targets applications in the domain of factory automation where determinism in the time domain is important. In a nutshell, the LLDN extension specifies a star network in which a number of sensors are associated with a coordinator, and where deterministic and low-latency transmission is achieved through a TDMA-like (Time-Division Multiple Access) medium access control scheme. In this scheme, time is partitioned into subsequent superframes, which, in turn, are partitioned into time slots, most of which are allocated for exclusive use to sensors (for uplink traffic) and to the coordinator (for downlink traffic). LLDN also addresses reliability concerns by including both individual and group acknowledgements and explicit retransmission slots.

In this paper, we explore how to use these retransmission slots to improve reliability for uplink traffic. According to the IEEE 802.15.4e amendment, each sensor gets one timeslot for an initial uplink transmission and at most one more timeslot for a retransmission (we call a source node whose initial transmission failed a **failed node** or a **failed source**). We argue that this allocation scheme is inappropriate, for the following reasons: (i) a failed source node can not use more than one retransmission slot, even if other retransmission slots are unused; and (ii) it ignores that the channels between different sources and the coordinator can be very different and can vary over time. We propose and evaluate the performance of a number of different schemes for allocating retransmission slots to failed nodes. One of the proposed schemes requires no changes to the IEEE 802.15.4e packet formats (only a behavioural change), other schemes only require modifications of one particular LLDN packet, the **group acknowledgement** (GACK). Most of our schemes are designed to maximize the probability that all source packets are eventually received within one superframe (we refer to this as the **success probability**). To achieve this, the coordinator continuously maintains estimates of the current packet error rates on the source-coordinator channels to inform retransmission slot allocation.

Our results indicate that it is possible to improve the success probability significantly already by simply allowing the system to use all available retransmission slots (and without requiring any changes to packet formats). However, much more substantial gains can be made by adding **cooperative relayers**, i.e., dedicated helper nodes tasked to overhear source transmissions and performing retransmissions on their behalf, leveraging spatial diversity [8]. It has already been shown in several other works that adding relayers to TDMA-based systems can substantially improve reliability (e.g., [9]). However, in many of these works, relay scheduling has been carried out under the assumption that all channels are known and time-invariant, or relayer selection is made instantaneously without exploiting what might have been learned about the efficacy of different relayers in the past. In this paper, we consider the situation where the channels are generally unknown and time-varying, and we propose a scheme in which the coordinator **learns** about the quality of different relayers to support the sources. The proposed scheme is based on algorithms to solve the multi-armed bandit problem, a well-studied problem in the field of reinforcement learning (RL) [10,11], and is designed to adapt to time-varying channels. Depending on the number of relayers, substantial improvements in the success probability can be achieved. We explore the performance characteristics of our learning-based scheme under a range of channel models and parameter settings.

The paper is structured as follows: in the next Section 2, we provide background information on LLDN and describe our system model. In Section 3, we introduce the retransmission scheduling schemes that do not employ relayers, and, in Section 4, we introduce the relayer-based retransmission schemes: one involving a practically feasible learning algorithm, and a genie-aided scheme for performance comparisons. All performance results have been obtained by simulation, and the simulation framework is explained in Section 5. The following Section 6 then contains our results. The learning scheme proposed in this paper uses one particularly important system parameter for generating actions, the so-called system temperature. In Section 6.1, we present simulation results justifying our choice of the system temperature value for the remaining paper. In Section 6.2, we compare the proposed schemes for unknown but static channels, in Section 6.3, we consider the case of time-varying channels, and in Section 6.4, we compare variants of the learning-based scheme with larger spaces of available actions against a genie-aided relaying scheme, and, through this, we are able to quantify the "system loss" from the operation of the learning scheme. Related work is summarized in Section 7 and our conclusions are given in Section 8.

This paper is a substantially extended and revised version of the conference contribution [12]. The main differences include the following: (i) the learning-based scheme has been generalized to allow for larger action spaces; (ii) another baseline scheme, the genie-aided scheme, has been added, allowing for assessment of the performance of the revised learning-based scheme in more detail; (iii) additional results exploring the choice of the system temperature parameter for the learning-based scheme and investigating the effect of a larger action space for the learning-based scheme have been added.

2. Background and System Model

2.1. Background on IEEE 802.15.4e LLDN

LLDN uses a TDMA-like medium access control scheme where time is partitioned into superframes. Two different superframe structures are specified for LLDN, which, for the purposes of this paper, are equivalent. We use the superframe structure shown in Figure 1. It starts with a beacon packet transmitted by the coordinator, followed by two optional management slots (e.g., for node association). Next, there are *K* uplink slots in which each of the *K* sensors has an initial transmission of its uplink data packet. In the following group acknowledgement (GACK) packet, the coordinator broadcasts a bitmap indicating the received uplink packets, using one bit for each source node. A configurable number *N* of retransmission uplink slots follows the GACK. According to the amendment, the first retransmission slot is allocated to the failed source coming first in the bitmap, the second slot is allocated to the failed node coming second, and so on. Hence, a failed source will not get more than one retransmission slot. Up to *N* retransmission slots are available, and if there are more than *N* failed sources, then the last ones will miss out. Lastly, there is optionally a number of so-called bi-directional slots that can be allocated to the coordinator or to sensor nodes. In this paper, however, we focus entirely on the uplink and retransmission slots and ignore the management and bi-directional slots.

In the second superframe format specified in the amendment, the group acknowledgement is part of the beacon packet and retransmission slots are placed at the beginning of the next superframe.

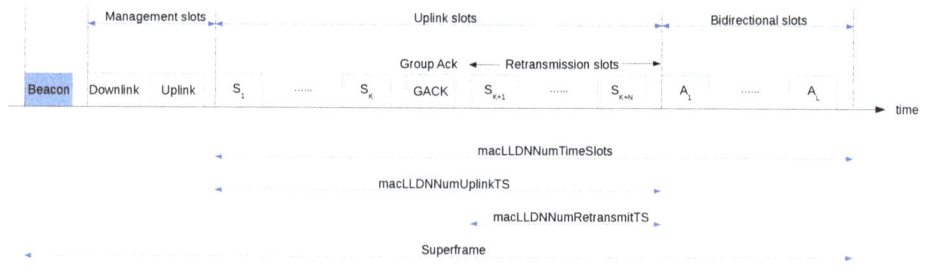

Figure 1. Superframe structure (with separate group acknowledgement), compare ([6] Figure 11g).

2.2. System Model

We look at a single LLDN network having a star topology. There is a single PAN (Personal Area Network) coordinator, a number *K* of sensor nodes, and *R* relay nodes. We assume that the relay nodes are truly distinct from the sensor nodes, but they can also be integrated with source nodes [13]. The network has already reached steady-state operation, i.e., all nodes are registered with the coordinator and have started transmitting data packets. There are *N* retransmission slots, no management slots and no bi-directional slots. To cover the worst case, we assume that each source node has a new packet at the beginning of each superframe, which needs to be successfully transmitted before the superframe ends.

To focus this study on the effects of the actual retransmission scheduling methods, we assume that all transmissions coming from the coordinator (beacons, GACK packets) are completely and reliably received by all other nodes. However, the other channels (uplink channels from sources and relays to the controller, channels between sources and relays) can introduce packet errors. We assume that these channels are pairwise statistically independent, and in order to simplify the channel model, we assume that all data packets have the same length, so that it is meaningful to assign packet error rates to individual channels. In applications where sensor nodes transmit sensor data, this assumption is realistic, as sensor readings are often very small and make up only a relatively small part of a packet, which needs to have physical layer preambles, headers, checksums and other overheads. There is no external interference in the system.

We use two channel models. In the **static channel model**, the packet error rate for each channel is drawn randomly from a uniform distribution between 0% and 100% and remains fixed throughout. In the **time-varying channel model**, each channel changes between two different packet error rates (each drawn randomly from a uniform distribution between 0% and 100%) following a two-state time-homogeneous Markov chain with two identical average state holding times.

3. Non-Relaying Schemes

In this section, we describe the retransmission slot allocation schemes that do not use relay nodes.

3.1. Standard-Based Schemes

The **standard scheme** (Std) implements the allocation of retransmission slots according to the IEEE 802.15.4e LLDN specification (see Section 2.1). Each failed source gets at most one retransmission slot (exactly one if there are at least as many retransmission slots as there are failed sources). When there are fewer failed sources than retransmission slots, some of the retransmission slots remain unused. The **enhanced standard scheme** (EnhStd) does not need any changes to the GACK packet format, but a change in the behaviour of sources and the controller. If we have N retransmission slots and sources $s_1, \ldots, s_m$ have failed (ordered according to their position in the GACK bitmap), the retransmission slots are allocated in a repeating cycle as $s_1, s_2, \ldots, s_m, s_1, s_2, \ldots, s_m, s_1, s_2, \ldots$ until all N slots are exhausted. Hence, sources can get more than one retransmission slot and all N retransmission slots are fully utilized.

3.2. The Optimal(PAR) Scheme

In this subsection and the next, we introduce two schemes that use an estimate of the current packet error rate (PER) for each source-controller channel. For each source node i, the controller maintains a PER estimate $p_i(t)$ that is updated in each superframe immediately after the initial source transmissions and before the allocation of retransmission slots and subsequent transmission of the GACK frame (the time index $t \in \mathbb{N}$ counts the superframes). If the controller did not receive the packet from source i, it encodes the outcome as $o_i(t) = 1$—otherwise (in case of successful reception) as $o_i(t) = 0$. The PER estimate is updated using an exponentially-weighted moving average (EWMA) algorithm:

$$p_i(t) = \alpha \cdot o_i(t) + (1 - \alpha) \cdot p_i(t - 1), \tag{1}$$

where $0 < \alpha < 1$ is a parameter and the initial values $p_i(0)$ are set to zero. This is a well-known method for PER estimation and has the ability to adapt to changing channels. In this paper, we assume throughout that $\alpha = 0.03$, i.e., most weight is put on the "history" summarized in $p_i(t-1)$.

In the **optimal(PAR) scheme** (OptPAR, PAR is a shorthand for "Probability that All packets are Received"), the controller considers all possible allocations of N slots to the M failed nodes, and calculates for each such allocation the probability that the controller receives all packets successfully (see Equation (3)). The controller retains the allocation which maximizes this measure. Due to its computational complexity, we do not consider this scheme as an option for practical implementation, since the number of such allocations is ([14] Section II.5):

$$\binom{M + N - 1}{N}. \tag{2}$$

However, we have included this scheme to compare its performance with that of the (much more practical) heuristic(PAR) scheme, which is discussed next.

3.3. The Heuristic(PAR) Scheme

In the **heuristic(PAR) scheme**, the current PER estimates $p_i(t)$ again play a role in allocating the N available retransmission slots to the M failed sources in the current superframe. For our presentation, we simply number the failed sources from 1 to M.

According to our assumptions, we have M statistically independent wireless links, and on each link i packet errors happen independently of each other with PER p_i (we drop the dependence on time for notational convenience). We are furthermore given N retransmission slots and we assume that $N > M$ holds. (For $N \leq M$, we simply assign one retransmission slot to each of the first N failed sources.). Suppose that station i gets allocated n_i retransmission slots, in which station i simply repeats its packets n_i times. (Note that, for simplicity, we ignore feedback from the controller here. In practice, when a source node gets positive feedback after fewer than n_i retransmissions, it may stop. However, this will only impact the energy consumption of the source node and not its reliability, which is the main focus of this paper.). Then, the probability that node i's packet fails to reach the controller is $p_i^{n_i}$, and the probability that the controller receives **all** packets becomes (by independence of channels/sources):

$$\Pr\left[\text{Success}\right] = \prod_{i=1}^{M} \left(1 - p_i^{n_i}\right). \tag{3}$$

Assuming that $p_1, \ldots, p_M$ are given, we want to pick numbers $n_1, \ldots, n_M \in \mathbb{N}$ that maximize this expression (equivalently: its logarithm) and which obey $n_1 + \ldots + n_M = N$ to make sure that all N slots are allocated. Hence, we get the following integer optimization problem:

$$\text{maximize} \quad f(n_1, \ldots, n_M) = \sum_{i=1}^{M} \log\left(1 - p_i^{n_i}\right), \tag{4}$$

$$\text{s.t.} \quad \sum_{i=1}^{M} n_i = N,$$

$$n_i \in \mathbb{N},$$

$$n_i \geq 1 \quad (i \in \{1, \ldots, M\}),$$

where the last condition ensures that each failed source gets at least one retransmission slot. Since integer optimization in general is NP-hard [15], we have developed a heuristic based on the relaxation of this problem (ignoring the constraint $n_i \geq 1$ for the time being) and the Lagrange multiplier method [16]. The Lagrangian of this problem is

$$L(n_1, \ldots, n_M, \lambda) = f(n_1, \ldots, n_M) + \lambda \left(\sum_{i=1}^{M} n_i - N \right), \tag{5}$$

where $\lambda \in \mathbb{R}$ is the Lagrange multiplier. To apply the Lagrange multiplier theorem ([16] Thm. 3.1.1), the partial derivatives of the Lagrangian are needed:

$$\frac{\partial}{\partial n_i} L(n_1, \ldots, n_M, \lambda) = \frac{-\log(p_i) \cdot p_i^{n_i}}{1 - p_i^{n_i}} + \lambda,$$

$$\frac{\partial}{\partial \lambda} L(n_1, \ldots, n_M, \lambda) = \sum_{i=1}^{M} n_i - N.$$

With the abbreviation $c_i = \log(p_i)$, the equation $\frac{\partial}{\partial n_i} L(\cdot) = 0$ can be solved for $n_i = n_i(\lambda)$ as:

$$n_i(\lambda) = \frac{\log\left(\frac{\lambda}{c_i + \lambda}\right)}{c_i}. \tag{6}$$

Plugging this expression for n_i into the second condition $\frac{\partial}{\partial \lambda} L(\cdot) = 0$ yields:

$$0 = \sum_{i=1}^{M} \frac{\log\left(\frac{\lambda}{c_i + \lambda}\right)}{c_i} - N. \tag{7}$$

This needs to be solved for λ to find the multiplier and subsequently the n_i (from Equation (6)). For $M > 1$, there is in general no closed-form expression for λ and we need to resort to numerical computation. Since the functions $n_i(\cdot)$ from Equation (6) should return positive values, the fact that $c_i < 0$ implies that $0 < \frac{\lambda}{c_i+\lambda} < 1$ must hold, which in turn requires $\lambda < 0$. Furthermore, noting that for $\lambda < 0$ the function $n_i(\cdot)$ is monotonically increasing, and using $\lim_{\lambda \to -\infty} n_i(\lambda) = 0$ and $\lim_{\lambda \to 0-} = \infty$, we can conclude that the right-hand side of Equation (7) has exactly one root λ^*, which can be found efficiently using a bisection method. Please note that solutions with $n_i(\lambda^*) < 1$ cannot be ruled out.

With these building blocks in place, we can now describe our heuristic(PAR) algorithm. Assume that there are M failed sources and N retransmission slots. The current PER estimates for the failed nodes are given by $p_1, \ldots, p_M$, and we first compute the Lagrange multiplier λ^* as the unique negative root of Equation (7). Then, set

$$n_i^* = \lfloor n_i(\lambda^*) \rfloor, \tag{8}$$

as our initial guess for the optimal $n_1, \ldots, n_M$. The function $\lfloor x \rfloor$ returns the largest integer $\le x$.

There can be cases where

$$N' := \sum_{i=1}^{M} n_i^* < N$$

holds and some slots are not used. We first allocate one of the unused slots to each source i with $n_i^* = 0$ while possible, and update their slot counters to $n_i^* = 1$. After doing this, when there are still $N'' = N - \sum_{i=1}^{M} n_i^* > 0$ unused slots, we run the following algorithm to allocate these N'' slots based on the "allocation gap":

```
while  N'' > 0:
    j := arg max_{i∈{1,...,M}} (n_i(λ*) − n_i*)
    allocate  slot  to  failed  source  j
    n_j* := n_j* + 1
    N'' := N'' − 1
```

Computationally, the most complex step in this algorithm is finding the root of a strictly monotonically increasing function (Equation (7)), which can be done efficiently to arbitrary precision using a bisection method. We surmise that the entire algorithm is quick enough so that the resulting allocation is available when the controller starts to construct the GACK packet. However, the GACK packet needs to be extended: beyond the acknowledgement bitmap, it needs to indicate for each failed source i the number n_i^* of retransmission slots it gets. When N is not too large, these numbers only need a few bits per failed source.

4. Relaying Schemes

In this section, we introduce the two relaying schemes considered in this paper. The first is the learning-based scheme, the second is an idealized genie-aided scheme introduced for comparison purposes.

4.1. Learning-Based Scheme

Diversity schemes, in particular spatial diversity schemes, are a key approach to improving transmission reliability over wireless channels [8,17]. In spatial diversity schemes, multiple spatially separated antennas are used to transmit or receive information. In the special case of cooperative communications, the required antennas are "borrowed" from third-party nodes, called **relayers** [18,19]. A particularly interesting application of the cooperative communications concept is to incorporate relay nodes into retransmission-based error-control schemes, where a relayer performs retransmissions on behalf of a source node, provided the relay node has managed to overhear the original data packet (see [9,20] for relaying in a TDMA context).

We assume that, besides the K source nodes, there are R separate relay nodes in the network. The relays are switched on all the time to overhear packets transmitted by source nodes. In our scheme, it is the controller which decides which relay will have to support a given source node. Note, however, that the controller has initially no information about the (generally time-varying) channels between the sources and relays, and between the relay nodes and the controller.

The decision scheme used to allocate a relayer to a failed source has some similarities to learning-based algorithms used to solve the multi-armed bandit problem, a well-known problem in reinforcement learning (RL) [10,11]. Broadly, the controller needs to balance two different goals: on the one hand, it should consistently apply the action (choosing a relay node and its number of slots) that is currently known to be the best one (this is called **exploitation** in the literature); on the other hand, it has to test other actions from time to time to see whether they give better results than previously thought and to update the knowledge about the quality of other actions (this is called **exploration**). Exploration is particularly important for time-varying channels. A standard approach is to switch probabilistically between exploration and exploitation, and, in this paper, we do this by using the Boltzmann distribution for action selection (see below).

Our learning-based scheme uses the heuristic(PAR) scheme as a starting point and we refer to it as the **learning(PAR) scheme**. Important design goals are simplicity (so it can be executed in real time) and quick convergence towards optimal actions for given channel conditions. In the light of the convergence requirement, we have decided to keep the space of available actions small; in particular, we allow only one relayer to support a failed source within a superframe and ignore the possibility to build elaborate relaying chains involving two or more distinct relayers.

The controller runs a separate instance of our algorithm for each source node, and, in the following, we consider a fixed failed source node i. After the initial transmissions, the controller first invokes the heuristic(PAR) scheme to calculate an initial allocation of retransmission slots to failed nodes (note that here the current channel PER estimates come in). For our failed node i, we denote by n_i the number of allocated retransmission slots, and regard this as its *state*. Therefore, the state space is given by the possible number of retransmission slots a failed node can get, which ranges from 1 to N.

In each state, the controller picks an action from an action set. An action a in state $s \in \{1, \ldots, N\}$ is given by a pair $a = (r, m)$ specifying a relayer r and a number m of contiguous slots given to the relayer. For $s = 1$, we require $m = 0$ and for $s > 1$, we restrict m to be from the set $\{1, \ldots, \min\{s - 1, \Delta\}\}$, where Δ is a protocol parameter specifying the maximum number of retransmission slots that can be allocated to a relayer. Note that the first retransmission slot is always allocated to the failed source i. When such an action is chosen, in the resulting slot allocation, the first $s - m$ slots are allocated to the failed source i, and the remaining m slots are allocated to relayer r (which is assumed to be only transmitting i's packet if it has overheard it previously, and otherwise remains silent). Besides the actions involving a relay node r, there is one further action in state s, which is to give all retransmission slots to the failed source i.

The controller stores for each source node i a separate table with all allowed state/action pairs. The table entries contain the average reward $Q(s, a)$ for state s and action a. Suppose that, in superframe t, the data packet of a particular source i has been received by the controller already after the initial transmissions. In this case, the table entries for this source remain unchanged. Otherwise, if source i is a failed source and has received s retransmission slots according to the heuristic(PAR) algorithm, the controller picks the action a for state s randomly, using a Boltzmann distribution (see below). After executing the s retransmission slots according to action a, the controller determines the outcome o: if the controller has eventually received the data packet of source i, it assigns $o = 1$ (indicating success); otherwise, it assigns $o = 0$. The value $Q(s, a)$ is then updated following an EWMA-type approach as

$$Q(s, a) := \alpha_r \cdot o + (1 - \alpha_r) \cdot Q(s, a), \tag{9}$$

where $0 < \alpha_r < 1$ is an adjustable parameter. Again, the EWMA-type reward update scheme helps with adaptation to time-varying channels. After a preliminary simulation-based performance evaluation, in this paper, we have fixed α_r as 0.05, i.e., most weight is on the "history".

There are different (randomized) methods to pick an action a in a given state s. In the conference version [12], we have used a scheme in which, with a fixed probability ϵ, the best available action for state s is chosen (i.e., the action a maximizing $Q(s,a)$), and, with probability $1 - \epsilon$, an action is chosen randomly with uniform distribution. However, we found that better performance could be achieved when we use a Boltzmann distribution with a given temperature parameter $\tau > 0$ to pick an action, and this is what we use in this paper. In this method, the action a for given state s is always chosen randomly according to the following distribution:

$$\Pr\left[\text{Action} = a^*\right] = \frac{\exp\left(\frac{Q(s,a^*)}{\tau}\right)}{\sum_a \exp\left(\frac{Q(s,a)}{\tau}\right)}, \tag{10}$$

where the sum in the denominator extends over all actions a available in state s. We will consider the choice of the temperature parameter τ in Section 6.1. Overall, we found that the Boltzmann method can converge much quicker towards a good-quality action and is better suited for time-varying channels.

To implement this scheme, the GACK packet needs to be extended. In particular, for each failed source, we need to indicate the number of allocated retransmission slots (as in the heuristic(PAR) method), the chosen relayer r and the number m of slots allocated to the relayer. When the number of relayers and the parameter Δ are kept small, a few bits will suffice to encode the chosen action. Furthermore, the protocol needs to be extended by methods allowing relays to register themselves with the controller.

4.2. Genie-Aided Scheme

In the learning(PAR) scheme, we have deliberately kept the action space small (e.g., by using only one relayer instead of two or more for the same packet, and by limiting the number of slots the relayer can get), which potentially reduces its performance. Furthermore, the controller does not have any information about the channel quality between the source node and the relayer.

To assess the impact of limiting the number of slots a relayer can get, we consider an idealized relaying scheme, called the **genie(PAR)** scheme. Again, we discuss the behaviour for an individual failed node. The controller first runs the heuristic(PAR) scheme to obtain an initial allocation based on the current PER estimates, returning a number n_i of retransmission slots allocated to failed node i. Then, the controller uses divine insight to obtain the current true PERs of **all** the channels in the system and then calculates for each relayer r and each possible split of the n_i retransmission slots between the source node i and the relayer r (such that the first slot is always allocated to i and the relayer gets contiguous slots at the end) the success probability as

$$\left(1 - p_{i,r}^{n_i - n_r}\right) \cdot \left(1 - p_{r,c}^{n_r}\right),$$

where n_r is the number of slots allocated to the relayer, $p_{i,r}$ is the true channel PER between source i and relayer r, and $p_{r,c}$ is the true channel PER between relayer r and the controller. The best such allocation is then selected.

4.3. Overheads

The retransmission schemes presented in this paper are aimed at resource-constrained platforms, and so it is prudent to briefly consider their computational, memory and energy overheads.

Let us first consider the case of simple sensor nodes. As compared to the standard scheme, they do not have to carry out any additional computations and only require very little additional memory to store the slot indices of the retransmission slots allocated to them. In terms of energy, for the heuristic(PAR) and learning(PAR) schemes the sensor nodes will have to process slightly larger GACK frames (as these now contain additional bits describing the retransmission slot allocation) and will possibly have to carry out more than one retransmission. Sensor nodes can sleep outside beacon frames, GACK frames and their own retransmission frames. Relay nodes, however, would need to be awake during all initial sensor transmissions and need to engage in retransmissions, if called upon.

The case of the coordinator is slightly more complex, but it is an often-made assumption that centralized coordinator nodes are less resource-constrained than simple sensor nodes. The computational overhead of the heuristic(PAR) scheme involves some modest amount of floating-point calculations: after each initial sensor transmission in a superframe, the coordinator will have to update the PER estimate for this sensor according to Equation (1)—note that this also requires some memory to store the PER estimates, one floating-point value per sensor. Furthermore, before sending the GACK packet, it has to calculate the retransmission slot allocation, which involves root-finding of a strictly monotonic function (Equation (7), the number of terms corresponds to the number of sensors), running through the algorithm filling the allocation gap (see Section 3.3) and the actual construction of the GACK packet. The computational overhead of the learning-based scheme consists of updating the Q-values (Equation (9)) and generating random actions following the Boltzmann distribution (Equation (10)), which both happen once per superframe and for each failed node. Furthermore, the coordinator will require for each sensor node a table space of $1 + R \sum_{s=2}^{N} \min\{s - 1, \Delta\} \leq 1 + R(N - 1)\Delta$ floating point values to store the Q-values.

5. Simulation Framework

All performance results in this paper have been generated with a custom simulation tool written in the Haskell programming language. To explain the rationale for this, recall that the main focus of this paper is to understand the performance of the retransmission slot scheduling schemes in isolation, and so we have deliberately chosen to exclude several other system aspects that could possibly confound the performance results, such as channel coding and modulation, hardware aspects, complex channel models using fading at sub-superframe timescales, etc. We wanted to keep our system model as simple as possible (but not simpler), and correspondingly have opted to create a comparatively small custom simulator and avoid the complexities of simulation frameworks like ns-3 or OMNet++.

When using static channels, on an individual channel, all packets are erroneous independently of each other and with the same probability (the packet error rate). For a given set of simulation parameters, we draw for each of the $(K + R + 1)^2$ channels its packet error rate (PER) from a uniform distribution and run the system for 40,000 superframes or rounds. After finishing these 40,000 superframes, we compute the main performance measures considered in this paper: the most important measure is the **success probability**, defined as the fraction of superframes where **all** source packets have been received by the coordinator (a number between 0 and 1). Occasionally, we also consider the fraction of source packets received by the coordinator (again between 0 and 1). This procedure is repeated 100,000 times for the same parameter settings, so we run 100,000 replications with different random instantiations of the channel error probabilities. The output statistics of these replications are averaged, and we report these averages. Due to the large number of 100,000 replications the confidence interval half-widths for the success probability at a 99% confidence level are all below 0.3%, and we do not show the confidence intervals in the figures.

For time-varying channels, we use a modified version of this procedure. Each channel in the system is modeled as a two-state (time-homogeneous) Markov chain [21] with states s_1 and s_2. The state can only change at the start of a superframe and remains constant until the next superframe. In state s_i, the channel behaves like a static channel with packet error rate e_i. The packet error rates e_1 and e_2 are chosen randomly from a uniform distribution at the start of a replication, and, in each replication, 40,000 superframes are again simulated. The channel state transition probability matrix is the same for all channels and has the form

$$\mathbf{P} = \begin{pmatrix} p & 1 - p \\ 1 - p & p \end{pmatrix} \tag{11}$$

with parameter p. It is well-known that the state holding times have a geometric distribution [21] and the average state holding time in either state is $\frac{1}{1-p}$. In this paper, we consider four different values for p, giving channels with different rates of change: by picking $p \in \{0.9, 0.99, 0.999, 0.999999\}$, the state changes on average every 10th, 100th, 1,000th, or 1,000,000th superframe, respectively. Clearly, for larger values of p, the channels become more stable, and we loosely refer to p as the channel stability. As before, a single replication extends over 40,000 superframes and we have used 100,000 replications in total (except for Section 6.1, where due to the large number of parameter combinations, we have restricted ourselves to 10,000 replications).

6. Results

6.1. Choosing the System Temperature for the Learning(PAR) Scheme

In our first study, we assess the impact of the temperature parameter τ governing the Boltzmann distribution for selecting actions (see Section 4.1) on the achievable performance of the learning(PAR) scheme, using the time-varying channel model (see Section 5). The main performance parameter considered here is the success probability.

We have chosen a scenario with $K = 6$ sources and $N = 9$ retransmission slots. The learning(PAR) scheme runs with a maximum of $\Delta = 1$ retransmission slots that can be allocated to a relayer (see Section 4.1). We have varied the system temperature τ in the set $\tau \in \{0.05, 0.075, 0.1, \ldots, 0.225, 0.25, 0.3, 0.35, 0.4, 0.45, 0.5, 0.6\}$ and the number of relayers as $R \in \{1, 2, 3, 4, 5\}$. The success probability results for the four considered values of the channel stability p (see Section 5) are shown in Figure 2a–d, respectively. It can be seen that the choice of the temperature value has a quite significant impact on the success probability, and the variation over different temperatures becomes larger as the number of relayers increases. Somewhat to our surprise, for all considered channel stability values p and relayer numbers R, the temperature $\tau = 0.1$ has displayed the optimal or close-to-optimal performance, so, in the remaining part of the paper, we will only consider the learning(PAR) scheme with a temperature value of $\tau = 0.1$.

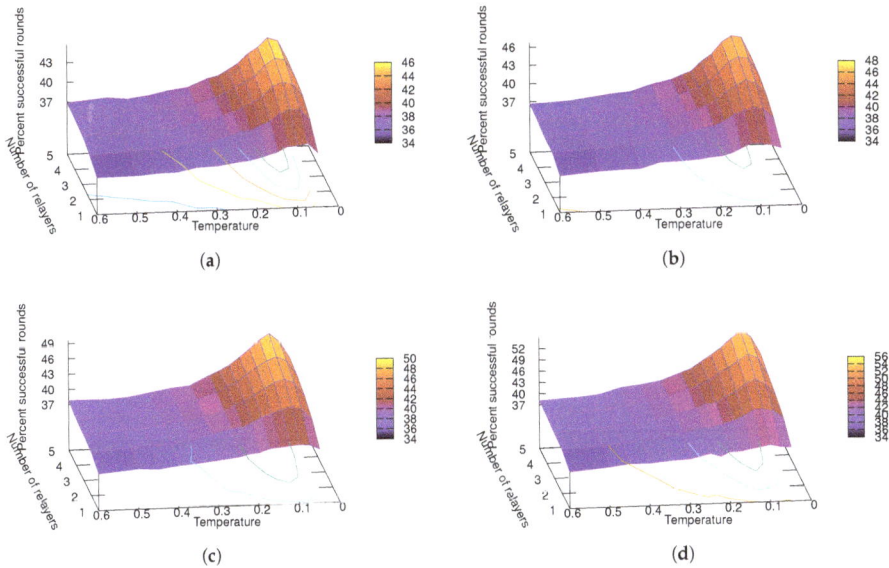

Figure 2. Average fraction of successful rounds for learning(PAR)-τ schemes for varying τ and varying numbers of relayers over two-state channels with different probabilities p to stay in the same state. (a) $p = 0.9$; (b) $p = 0.99$; (c) $p = 0.999$; (d) $p = 0.999999$.

6.2. Performance on Static Channels

In this section, we present simulation results for the case where all channels in the system follow the static channel model described in Section 5.

We have generated results for all the non-relaying schemes and the learning(PAR) scheme with a temperature parameter of $\tau = 0.1$ and, by setting $\Delta = 1$, restricting the relayer to a maximum of one slot. We have considered three different deployments:

- $K = 8$ source nodes, $N = 12$ retransmission slots,
- $K = 6$ source nodes, $N = 9$ retransmission slots,
- $K = 4$ source nodes, $N = 6$ retransmission slots.

In all of these deployments, the ratio of retransmission slots to source nodes is the same, giving each source 1.5 retransmission slots on average. For the learning(PAR) scheme, we have varied the number of relayers between 1 and 5.

In Figure 3a, we show the results for the success probability, and the fraction of received packets is displayed in Figure 3b The following points are interesting:

- The results for the success probability show a much wider spread (both when varying the number of sources and among the different schemes) than the fraction of successful packets. For the non-relaying schemes (except the standard scheme), the difference in the average fraction of successful packets is small, and the increase of that fraction for increasing numbers of relayers is moderate. Similar findings apply for all of the other scenarios studied in this paper, and we will not report further results on the fraction of successful packets.
- The standard scheme shows consistently and by some margin the poorest success probability performance. By comparing the standard scheme with the enhanced standard scheme, we can conclude that not utilizing all available retransmission slots significantly reduces the success probability.
- The success probability achieved by the heuristic(PAR) and optimal(PAR) schemes is very close, confirming that the heuristic proposed in Section 3.3 gives a very good approximation to the true optimum.
- Somewhat to our surprise, the heuristic(PAR) scheme shows almost the same success probability performance as the enhanced standard scheme—for six and eight sources, the advantage of heuristic(PAR) over the enhanced standard scheme is only on the order of 1% to 1.5% in absolute percentages.
- The biggest improvements can be achieved with the learning(PAR) scheme, in particular as more relayers are added to the system. For $K = 8$ sources and $R = 5$ relayers, the learning(PAR) scheme achieves almost twice the success probability of the heuristic(PAR) scheme; for smaller numbers of source nodes, the relative advantage is smaller but still significant. These results are even more encouraging when noting that the channels are completely random—with a carefully planned deployment of relayers further performance, improvements can be expected.

In summary, when only considering the non-relaying schemes, the enhanced standard scheme achieves almost the same performance as the heuristic(PAR) and optimal(PAR) schemes, while not requiring any changes to the LLDN packet formats. However, adding relayers to the system can achieve much more substantial gains.

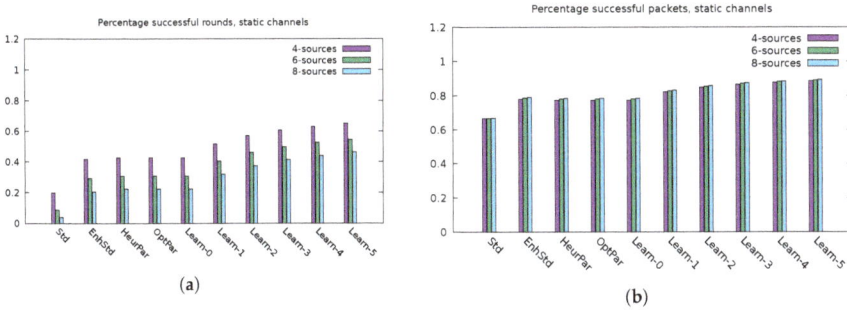

Figure 3. Average fractions of successful rounds and successful packets for all non-relaying schemes and learning (PAR). (**a**) success probability/fraction of successful rounds; (**b**) fraction of successful packets.

6.3. Performance on Time-Varying Channels

We next explore the success probability performance over time-varying channels. We have chosen a scenario with $K = 6$ sources and $N = 9$ retransmission slots, comparable with one of the scenarios discussed in Section 6.2 for static channels.

In Figure 4, we compare the success probability for all non-relaying schemes and the learning(PAR) scheme with a system temperature $\tau = 0.1$ and a maximum of $\Delta = 1$ retransmission slots allocated to a relayer. We present results for both static channels and the time-varying channels with the four different values for p, the channel stability (compare Section 5). A number of interesting observations can be made:

- The standard and enhanced standard schemes show more or less no sensitivity to the channel stability. The other two non-relaying schemes (heuristic(PAR), optimal(PAR)) show light performance improvements as the channel stability increases. We attribute this to the time required for the EWMA-based PER estimator (Equation (1)) after a channel change to adapt to the new channel PER. During this transient adaptation phase, sub-optimal allocation decisions can be made.

- When compared to static channels, the learning(PAR) scheme shows a reduced success probability performance over time-varying channels, particularly for smaller channel stability values. When the channel stability value becomes larger, the success probability of the learning(PAR) scheme approaches that for static channels, since, for larger channel stability values, the channels remain stable longer, and the fraction of time spent by the learning(PAR) scheme to learn the new channels becomes relatively smaller.

- Despite the performance loss observed over time-varying channels, it is still true for the learning(PAR) scheme that adding relayers gives significant success probability gains over the non-relaying schemes.

In summary, the learning(PAR) scheme can adapt to changing channels while successfully exploiting the presence of relayers. Higher channel stability leads to better performance for the learning schemes.

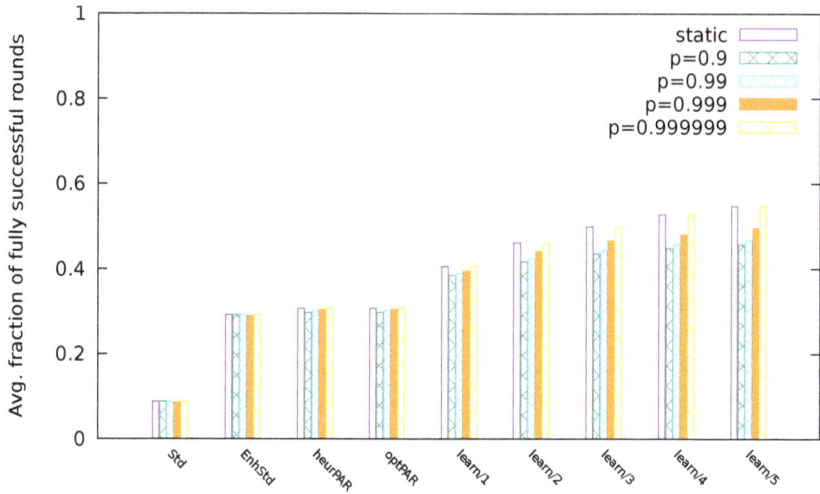

Figure 4. Average fraction of successful rounds.

6.4. Enlarging the Action Space

In the final experiment, we wanted to get some deeper insight into the performance characteristics of the learning(PAR) scheme. We have considered the learning(PAR) scheme for different numbers Δ of retransmission slots that can be allocated to a relay ($\Delta \in \{1, 2, 3, 4\}$—we denote the resulting scheme as learningPAR(Δ)) and compared it against the genie(PAR) scheme, both over the static channel model and the time-varying channel model. We have used a deployment with $K = 6$ sources, $N = 9$ retransmission slots and $R = 3$ relay nodes. Note that increasing Δ for the learning(PAR) scheme enlarges the space of available actions in a given state (compare Section 4.1), and it will on average require a longer time for the average rewards of all actions to settle close to their new values. The results are shown in Figure 5. The following points are interesting:

- Extending the action space for the learning(PAR) scheme has diminishing returns beyond $\Delta = 2$ for all considered channel models. The improvement from $\Delta = 1$ to $\Delta = 2$ is visible (most for the case of static channels), but, beyond this, it becomes marginal.
- In the case of static channels (and the time-varying channel with the largest channel stability), there is still a noticeable performance gap between the best learningPAR(Δ) scheme and the genie(PAR) scheme. We suspect that this gap is the price paid for the process of exploration, i.e., for the Boltzmann-based action selection scheme not selecting the best available action (which will have been learned after some time) throughout, but only with higher probability than other actions. Another possible explanation could have been the limited size of the action space when compared to the genie(PAR) scheme, but our finding of diminishing returns for increasing Δ does not support this hypothesis.
- The performance gap between the best learningPAR(Δ) scheme and the genie(PAR) scheme is even larger for time-varying channels with lower channel stability. The additional performance losses compared to static channels can be attributed to the transient times where the learningPAR(Δ) scheme needs to adjust to changed channels.

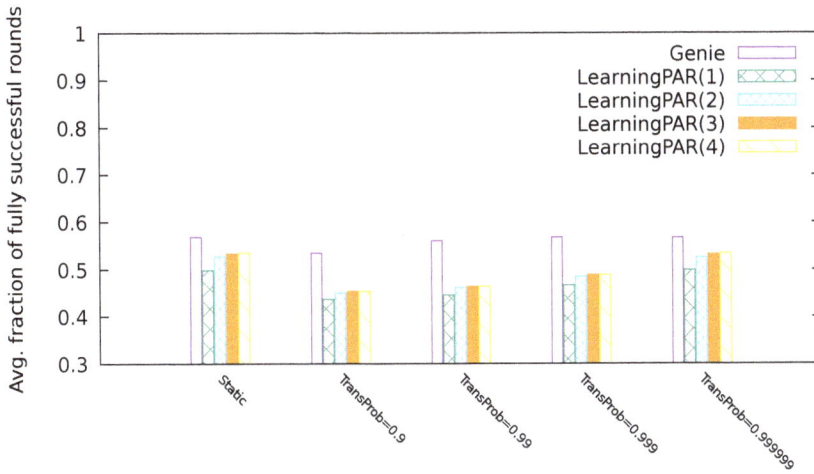

Figure 5. Average fraction of successful rounds.

7. Related Work

In the last few years, relay-assisted retransmission protocols (also known as cooperative ARQ protocols [22]) have received some interest in the area of industrial wireless (sensor) networks. The problem of optimal scheduling (with respect to the average number of packets successfully received before their deadline for a number of periodic sources with different periods) of relayer-based retransmissions has been explored in [9] for a TDMA-based system running over static (and known!) channels, and both (computationally heavy) optimal solutions and approximation algorithms have been investigated. The authors of [23] have considered relay-assisted communications in mining vehicle safety applications. Particularly, transmission scheduling for a multi-user MIMO system, including a set of relays, has been discussed. The application of relaying approaches in the context of IEEE 802.15.4 has been considered in [24], with a focus on relay selection. Three different selection schemes have been introduced, one in which relays are chosen periodically, one in which relay selection is triggered when the loss rate exceeds a threshold, and one in which each failed packet triggers a new relay selection. These schemes have been assessed experimentally for their effectiveness and their coordination overhead. All the proposed schemes require new control frames and involve explicit signaling. Note that the learning(PAR) scheme proposed in this paper differs from the schemes proposed in [24] by constantly maintaining quality information about all relayers, so that, after a degradation of the relayer that was best so far, the next best relayer is known quickly. The authors of [25] discuss the issue of cooperative relayer selection in an IEEE 802.15.4e variant. In their approach, the relayer for a given source node is selected only once and then not changed afterwards.

The LLDN extension to IEEE 802.15.4 has also been the focus of other works. The authors of [26] propose a method incorporating a time-diversity scheme to improve reliability. In this scheme, a packet is repeated a number of times in a superframe, and the impact on the (application-level) control performance of a networked control system is investigated. In [27], the performance of LLDN is analyzed in some detail; in particular, overheads and latencies are carefully explored.

The IEEE 802.15.4e amendment contains further extensions beyond LLDN. One of them, the time-slotted channel-hopping (TSCH) extension where the system periodically changes its frequency channel, has been considered in [28]. The link reliability of IEEE 802.15.4 in an industrial environment has been assessed experimentally in [29].

In summary,, in this paper, we have introduced and investigated a novel retransmission- and relay scheduling scheme based on Q-learning, which constantly keeps track of channel quality information and can identify new relayers quickly after channel changes. We have also carried out a detailed performance investigation.

8. Conclusions

The results in this paper suggest that the retransmission scheduling method described in the LLDN extension is sub-optimal and can be improved substantially. A substantial improvement can already be made when simply allowing the system to make use of all available retransmission slots and giving more than one slot to a failed source —as discussed, the enhanced standard scheme does not even require any changes to packet formats and is almost as good (in the considered scenarios) as more optimized non-relaying schemes. Significant further gains can be achieved by adding relayers, at the cost of changing the GACK packet format, some additional signaling to announce the presence of relay nodes to the controller, and the introduction of some additional configuration values (α, α_r, Δ and the temperature parameter τ). However, our learning-based scheme does **not** require any a priori channel state information and can adapt to changing channels. The price paid for the ability to adapt can be quantified by the performance difference between the learningPAR(Δ) scheme and the genie(PAR) scheme.

There are several avenues for future work. One is the more systematic offline (or perhaps even online) optimization of the system parameters (α, α_r, Δ, τ). For example, the two parameters α and α_r determining the "learning speed" for the channel PER estimator and the rewards can be adapted according to the observed rate of change in the underlying wireless channels. When this rate of change becomes small, one may also consider "cooling down" the system temperature (i.e., making τ smaller) to more strongly prefer the best available action. Furthermore, it would be very interesting to implement the learning(PAR) scheme in an experimental testbed to obtain better insight into implementation issues and to perform performance measurements. Finally, to eliminate the need for dedicated relay nodes, one could also extend the learning(PAR) scheme to use source nodes as relayers, energy consumption concerns permitting.

Acknowledgments: We would like to thank the anonymous reviewers and the editor for their helpful and constructive comments.

Author Contributions: Andreas Willig contributed to algorithm and conceptual design, conducted the simulations, evaluated the results and wrote the article. Yakir Matusovsky and Adriel Kind contributed to algorithm and conceptual design.

Conflicts of Interest: The authors declare no conflict of interest.

References

1. Galloway, B.; Hancke, G.P. Introduction to Industrial Control Networks. *IEEE Commun. Surv. Tutor.* **2013**, *15*, 860–880.
2. Gungor, V.C.; Hancke, G.P. Industrial Wireless Sensor Networks: Challenges, Design Principles, and Technical Approaches. *IEEE Trans. Ind. Electron.* **2009**, *56*, 4258–4265.
3. Willig, A. Recent and Emerging Topics in Wireless Industrial Communications: A Selection. *IEEE Trans. Ind. Inform.* **2008**, *4*, 102–124.
4. LAN/MAN Standards Committee of the IEEE Computer Society. *IEEE Standard for Local and Metropolitan Area Networks—Part 15.4: Low Rate Wireless Personal Area Networks (LR-WPANs)*; Revision of 2011; Institute of Electrical and Electronics Engineers: Piscataway, NJ, USA, 2011.
5. Toscano, E.; Bello, L.L. Multichannel Superframe Scheduling for IEEE 802.15.4 Industrial Wireless Sensor Networks. *IEEE Trans. Ind. Inform.* **2012**, *8*, 337–350.
6. IEEE Computer Society. *IEEE Standard for Local and Metropolitan Area Networks–Part 15.4: Low-Rate Wireless Personal Area Networks (LR-WPANs)— Amendment 1: MAC Sublayer*; IEEE Std 802.15.4e-2012; Institute of Electrical and Electronics Engineers: Piscataway, NJ, USA, 2012.

7. Chen, F.; German, R.; Dressler, F. Towards IEEE 802.15.4e: A Study of Performance Aspects. In Proceedings of the 8th IEEE International Conference on Pervasive Computing and Communications (PERCOM Workshops), Mannheim, Germany, 29 March–2 April 2010; pp. 68–73.
8. Diggavi, S.N.; Al-Dhahir, N.; Stamoulis, A.; Calderbank, A.R. Great Expectations: The Value of Spatial Diversity in Wireless Networks. *IEEE Proc.* **2004**, *92*, 219–270.
9. Willig, A.; Uhlemann, E. Deadline-Aware Scheduling of Cooperative Relayers in TDMA-Based Wireless Industrial Networks. *Wirel. Netw.* **2014**, *20*, 73–88.
10. Sutton, R.S.; Barto, A.G. *Reinforcement Learning—An Introduction*; MIT Press: Cambridge, MA, USA, 1998.
11. Bertsekas, D.P.; Tsitsiklis, J.N. *Neuro—Dynamic Programming*; Athena Scientific: Belmont, MA, USA, 1996.
12. Willig, A.; Matusovsky, Y.; Kind, A. Retransmission Scheduling in 802.15.4e LLDN—A Reinforcement Learning Approach with Relayers. In Proceedings of the International Telecommunication Networks and Applications Conference (ITNAC) 2016, Dunedin, New Zealand, 7–9 December 2016.
13. Girs, S.; Willig, A.; Uhlemann, E.; Bjoerkman, M. Scheduling for Source Relaying with Packet Aggregation in Industrial Wireless Networks. *IEEE Trans. Ind. Inform.* **2016**, *12*, 1855–1864.
14. Feller, W. *An Introduction to Probability Theory and Its Applications—Volume I*, 3rd ed.; John Wiley: New York, NY, USA, 1968.
15. Korte, B.; Vygen, J. *Combinatorial Optimization—Theory and Algorithms*, 3rd ed.; Springer: Berlin, Germany, 2005.
16. Bertsekas, D.P. *Nonlinear Programming*, 2nd ed.; Athena Scientific: Belmont, MA, USA, 1999.
17. Rappaport, T.S. *Wireless Communications—Principles and Practice*; Prentice Hall: Upper Saddle River, NJ, USA, 2002.
18. Liu, K.J.R.; Sadek, A.K.; Su, W.; Kwasinski, A. *Cooperative Communications and Networking*; Cambridge University Press: Cambridge, UK, 2009.
19. Kramer, G.; Maric, I.; Yates, R.D. Cooperative Communications. *Found. Trends Netw.* **2006**, *1*, 271–425.
20. Laneman, J.N.; Tse, D.N.C.; Wornell, G.W. Cooperative Diversity in Wireless Networks: Efficient Protocols and Outage Behaviour. *IEEE Trans. Inf. Theory* **2004**, *50*, 3062–3080.
21. Norris, J.R. *Markov Chains*; Cambridge University Press: Cambridge, UK, 1997.
22. Marchenko, N.; Bettstetter, C. Cooperative ARQ with Relay Selection: An Analytical Framework Using Semi-Markov Processes. *IEEE Trans. Veh. Technol.* **2014**, *63*, 178–190.
23. Ni, W.; Collings, I.B.; Liu, R.P.; Chen, Z. Relay-Assisted Wireless Communication Systems in Mining Vehicle Safety Applications. *IEEE Trans. Ind. Inform.* **2014**, *10*, 615–627.
24. Marchenko, N.; Andre, T.; Brandner, G.; Masood, W.; Bettstetter, C. An Experimental Study of Selective Cooperative Relaying in Industrial Wireless Sensor Networks. *IEEE Trans. Ind. Inform.* **2014**, *10*, 1806–1816.
25. Momoda, M.; Hara, S. Use of IEEE 802.15.4 for a Cooperator-Assisted Wireless Body Area Network. In Proceedings of the 8th International Symposium on Medical Information and Communication Technology (ISMICT), Firenze, Italy, 2–4 April 2014; pp. 1–5.
26. Yen, B.X.; Hop, D.T.; Yoo, M. Redundant Transmission in wireless networked control system over IEEE 802.15.4e. In Proceedings of the International Conference on Information Networking (ICOIN), Bangkok, Thailand, 28–30 January 2013; pp. 628–631.
27. Dariz, L.; Malaguti, G.; Ruggeri, M. Performance Analysis of IEEE 802.15.4 real-time Enhancement. In Proceedings of the IEEE 23rd International Symposium on Industrial Electronics (ISIE), Harbiye Istanbul, Turkey, 1–4 June 2014; pp. 1475–1480.
28. Palattella, M.R.; Accettura, N.; Grieco, L.A.; Boggia, G.; Dohler, M.; Engel, T. On Optimal Scheduling in Duty-Cycled Industrial IoT Applications Using IEEE802.15.4e TSCH. *IEEE Sens. J.* **2013**, *13*, 3655–3666.
29. Yadong, W.; Shihong, D. Study on IEEE802.15.4 Link Reliability in Industrial Environments. In Proceedings of the IEEE International Conference on Wireless for Space and Extreme Environments (WiSEE), Baltimore, MD, USA, 7–9 November 2013; pp. 1–7.

Journal of
*Sensor and
Actuator Networks*

MDPI

Article

An Experimental Evaluation of the Reliability of LoRa Long-Range Low-Power Wireless Communication

Marco Cattani *, Carlo Alberto Boano and Kay Römer

Institute for Technical Informatics, Graz University of Technology, Graz 8010, Austria;
cboano@tugraz.at (C.A.B.); roemer@tugraz.at (K.R.)
* Correspondence: m.cattani@tugraz.at; Tel.: +43-316-873-6910

Received: 22 May 2017; Accepted: 9 June 2017; Published: 15 June 2017

Abstract: Recent technological innovations allow compact radios to transmit over long distances with minimal energy consumption and could drastically affect the way Internet of Things (IoT) technologies communicate in the near future. By extending the communication range of links, it is indeed possible to reduce the network diameter to a point that each node can communicate with almost every other node in the network directly. This drastically simplifies communication, removing the need of routing, and significantly reduces the overhead of data collection. Long-range low-power wireless technology, however, is still at its infancy, and it is yet unclear (i) whether it is sufficiently reliable to complement existing short-range and cellular technologies and (ii) which radio settings can sustain a high delivery rate while maximizing energy-efficiency. To shed light on this matter, this paper presents an extensive experimental study of the reliability of LoRa , one of the most promising long-range low-power wireless technologies to date. We focus our evaluation on the impact of physical layer settings on the effective data rate and energy efficiency of communications. Our results show that it is often not worth tuning parameters, thereby reducing the data rate in order to maximize the probability of successful reception, especially on links at the edge of their communication range. Furthermore, we study the impact of environmental factors on the performance of LoRa, and show that higher temperatures significantly decrease the received signal strength and may drastically affect packet reception.

Keywords: LoRa; long-range technology; environmental impact; temperature; link quality; outdoor; underground; indoor; energy-efficiency; reliability

1. Introduction

An increasing number of radio technologies enabling low-power wireless communication over long distances has emerged in the past years. Ultra-narrowband technologies such as Sigfox (Labège, France) and Weightless-N [1] (Cambridge, UK), as well as spread-spectrum technologies such as LoRa [2] (San Ramon, CA, USA), allow for communicating up to few kilometers, and to build up low-power wide area networks (LPWANs) that do not require the construction and maintenance of complex multi-hop topologies [3,4].

A key characteristic of LPWAN technologies is indeed the ability to *trade* throughput for range and vice versa, i.e., one has the ability to fine-tune physical layer (PHY) settings to select a more sensitive (but slow) configuration that allows communication over a longer distance. This flexibility makes LPWAN technologies particularly appealing to developers of Internet of Things (IoT) applications requiring long-range communications with relatively low data rates. At the same time, however, the ability to fine-tune PHY settings requires *a thorough understanding* of their impact on network performance, especially on the reliability and energy-efficiency of communications [5].

The research community has recently devoted significant attention to the role of PHY settings in the context of LPWANs [5–7], especially LoRa technology. Out of the existing LPWAN technologies,

LoRa has especially attracted a large body of work due to the availability of commercial off-the-shelf radio transceiver and platforms [8–10], as well as its ability to operate in an infrastructure-free manner and to build up ad hoc mesh networks [5]. LoRa-based networks have been deployed in several settings, ranging from indoor [4] and urban [7] environments, to maritime [11] and mountain scenarios [12]. These deployments have shown the impact of PHY settings on connectivity range and sensitivity [5,13], as well as having given a first impression of the packet reception ratio that can be achieved at different distances with different hardware platforms and physical layer configurations. Bor et al. [14] have also shown through simulation that the choice of the PHY settings affects the number of LoRa nodes that can concurrently access the channel, which has an impact on the scalability of LoRa networks. Furthermore, Bor and Roedig [4] have presented the results of systematic indoor experiments showing that the set of LoRa settings leading to the most energy-efficient operation dynamically changes over time. Based on these results, the authors proposed a protocol that periodically probes different settings and that dynamically picks the ones minimizing energy consumption at run-time.

Interplay between PHY settings and link quality. Although the aforementioned works started to shed light on how to carry out an optimal selection of LoRa's PHY settings, they all share a common assumption: the best performance is obtained in the presence of highly reliable links. Most works, indeed, specifically target PHY settings maximizing the link quality, i.e., focus on selecting physical layer configurations that allow to sustain a packet reception ratio of 90% or higher [4,12,14] This practice is likely influenced from the behavior of non opportunistic low-power wireless data collection protocols for IEEE 802.15.4 radios, which favor high-quality links to intermediate and lossy ones [15,16]. However, adjusting the PHY settings of the radio to maximize the link quality has important implications w.r.t. energy efficiency when using long-range low-power wireless technologies such as LoRa. Maximizing the link quality, indeed, typically implies an increase in the transmission power and data overhead, and the selection of a more sensitive (and hence slow) physical layer configuration. As a result, one increases not only the likelihood to receive packets, but also the energy consumption of the radio, due to the higher transmission power, and the radio-on time, due to larger PHY layer overhead. This observation raises a yet unanswered question: *is it worth selecting PHY settings to reduce the data rate in order to increase the link quality?* This question is particularly relevant when two nodes are at the edge of the communication range: should one select a setting that reduces the data rate to increase the robustness of communication (and aim for a link achieving a high packet reception ratio) or rather accept having a link of intermediate quality (i.e., experiencing some packet loss), but with high data rate, and implement a re-transmission scheme on top? How this choice affects the energy-efficiency of the network still needs to be investigated.

Impact of environmental conditions on communication performance. The characteristics of LPWANs make them suitable for outdoor deployments on a large scale, and it is hence important to study in detail the impact of *environmental effects* such as changes in meteorological conditions, as well as variations in temperature and humidity on network performance. Unfortunately, to date, there is still little understanding about the impact of the environment on the reliability of LoRa communication, especially for links that are at the edge of their communication range. Iova et al. [12] have reported the vulnerability of LoRa communications to environmental factors such as presence of vegetation and temperature variations, but without quantifying their impact. Other works in the low-power wireless community have shown that some IEEE 802.15.4 radios are particularly vulnerable to changes in temperature, and that even the daily fluctuations recorded outdoors can render a good link useless [17–20]. However, these results are platform-specific and cannot be generalized to LoRa transceivers. Therefore, *if and how much temperature affects LoRa's communication performance* is yet to be answered.

Our contributions. In this paper, we carry out an experimental evaluation of the reliability of LoRa in different settings and provide an answer to the aforementioned open questions. First, we study how PHY settings and environmental factors affect the reliability of LoRa communications through an extensive experimental campaign indoor, outdoor, and underground. In line with earlier works [7], our experiments show that PHY settings have a significant impact on packet reception rate and that indoor environments are more challenging for LoRa communications. Our results also suggest that it is better to use faster (but more fragile) settings together with a re-transmission mechanism rather then selecting resilient and slower settings, maximizing packet reception rate and link quality. A detailed study of the overhead of each PHY setting in relation to its improvement on packet reception rate indeed shows that setting a maximizing data rate and minimizing range should be preferred.

Furthermore, our experimental results show a clear correlation between temperature, humidity, packet reception rate, and received signal strength. We hence analyze in depth how environmental factors such as temperature variations affect the reliability of LoRa communications by performing a series of systematic experiments in controlled settings on different hardware platforms. These experiments show that the reliability of LoRa drastically decreases at high temperatures. On the one hand, the signal strength of received packets decreases linearly when temperature increases, as was also observed for a number of IEEE 802.15.4 radios [17,18]. On the other hand, the decrease in signal strength can significantly affect LoRa links that are at the edge of the communication range, increasing packet corruption and loss up to a point in which a link is totally compromised.

The contributions of this paper are hence threefold:

- We study how PHY settings and environmental factors affect the reliability of LoRa through an extensive experimental campaign indoor, outdoor, and underground;
- We analyze the impact of LoRa's PHY settings on the effective data rate and energy efficiency of communications, highlighting that it is not worth selecting settings to reduce the data rate in order to increase the link quality;
- We systematically study the impact of temperature on the reliability of LoRa communications and show that high temperatures decrease the received signal strength and drastically increase packet loss and corruption for nodes at the edge of the communication range.

The paper proceeds as follows. In the next section, we introduce the reader to long-range technologies and to the LoRa physical layer settings that can be configured to fine-tune the operations of LoRa transceivers. Section 3 highlights the yet open questions with respect to LoRa's reliability as a function of PHY settings and environmental conditions. In Section 4, we describe our experiments indoor, outdoor, and underground, highlighting the strong impact of the chosen PHY settings and environmental conditions on the reliability of communications. Thereafter, we investigate in detail the interplay between PHY settings and link quality in Section 5 and carry out experiments in controlled settings to quantify the impact of temperature on LoRa's communication performance in Section 6. We finally summarize our contributions in Section 7, along with a discussion of future work.

2. Primer on LPWANs and LoRa

Low-power wide area networks complement short range wireless technologies such as Wi-Fi, Bluetooth Low Energy, and IEEE 802.15.4, and represent an interesting alternative to cellular technologies for urban-scale IoT applications. The success of LPWANs is due to their ability of providing long-range communication to thousands of devices at minimal cost and limited energy expenditure. Longer communication ranges allow for drastically simplifying duty cycling and networking protocol, as LPWANs can form star topologies where the low-power end devices are able to directly communicate with a more powerful orchestrator. This also allows for designing asymmetric communication schemes and to shift the load to the more powerful central device.

In order to increase the communication range, LPWAN technologies must improve the signal-to-noise ratio (SNR) at the receiver, either by narrowing down the receiver's bandwidth (reducing the receiver's noise-floor) or by spreading the energy of the signal over a wider freuency band (effectively reducing the spectral power density of the signal) [5]. NB-IoT [21] and Weightless-P [22], for example, encode the signal in low bandwidth (<25 kHz) to reduce the noise level and keep the transceiver design as simple and cheap as possible. Sigfox [23] and Weightless-N [24] further narrow the signal into ultra-narrow bands as narrow as 100 Hz, further reducing the perceived noise.

LoRa technology. Compared to these technologies, LoRa spreads the signal over a wider frequency band, and is more resilient to jamming and interference. LoRa is a proprietary LPWAN technology from Semtech (Camarillo, CA, USA) that recently attracted significant attention due to its ability to trade efficiently communication range against high data-rates, thus enabling IoT applications at an urban scale. The core of LoRa technology is its Chirp Spread Spectrum (CSS) modulation: the carrier signal of LoRa consists of *chirps*, signals whose frequency increases or decreases over time. LoRa's chirps allow the signal to travel long distances and to be demodulated even when its power is up to 20 dB lower than the noise floor. Because of this aspect, carrier sensing in LoRa is quite challenging: LoRa radios allow carrier detection via a *CAD mode*, a special reception state consuming half of the energy compared to the normal reception mode. However, the signals produced by different LoRa networks operating on different settings could create interference leading to false detections [7].

LoRa's communication performance can be fine-tuned by varying the selection of several PHY settings, including bandwidth, spreading factor, coding rate, transmission power, and carrier frequency, as summarized in Table 1. We explain next in detail the impact of each PHY parameters on data rate, receiver sensitivity (including resilience to interference), transmission range, and energy-efficiency [25].

Table 1. Summary of LoRa's configurable settings and their impact on communication performance.

Setting	Values	Effects
Bandwidth	$125\ldots500\,$kHz	Higher bandwidths allow for transmitting packets at higher data rates (1 kHz = 1 kcps), but reduce receiver sensitivity and communication range.
Spreading Factor	$2^6\ldots2^{12}\,\frac{\text{chips}}{\text{symbol}}$	Bigger spreading factors increase the signal-to-noise ratio and hence radio sensitivity, augmenting the communication range at the cost of longer packets and hence a higher energy expenditure.
Coding Rate	$4/5\ldots4/8$	Larger coding rates increase the resilience to interference bursts and decoding errors at the cost of longer packets and a higher energy expenditure.
Transmission Power	$-4\ldots20\,$dBm	Higher transmission powers reduce the signal-to-noise ratio at the cost of an increase in the energy consumption of the transmitter.

Bandwidth (BW). Varying the range of frequencies (bandwidth) over which LoRa chirp spread allows for trading radio air time against radio sensitivity, thus energy efficiency against communication range and robustness. The higher is the bandwidth, the shorter is the air time and the lower is the sensitivity. A lower bandwidth also requires a more accurate crystal in order to minimize problems related to the clock drift. Given a bandwidth BW, typically in the range of $125\ldots500\,$kHz, LoRa's chip-rate R_C is computed as:

$$R_C - BW \quad chips/s.$$

Spreading Factor (SF). To transmit information, LoRa "spreads" each symbol over several chips (*spreading factor*) to increase the receiver's sensitivity even more. LoRa's spreading factor SF can be selected between 6 and 12, resulting in a spreading rate ranging from 2^6 to 2^{12} chips/symbol and a symbol-rate R_S that can be computed as:

$$R_S = \frac{R_C}{2^{SF}} = \frac{BW}{2^{SF}} \quad symbols/s,$$

and resulting in a modulation bit-rate that can be expressed as:

$$R_M = SF \cdot R_S = SF \cdot \frac{BW}{2^{SF}} \quad bits/s.$$

Note that, in LoRa, packets transmitted with different spreading factors are orthogonal with each other and do not cause collisions if transmitted concurrently.

Coding Rate (CR). To increase the resilience to corrupted bits, LoRa supports forward error correction techniques with a variable number CR of redundant bits, ranging from 1 to 4. The resulting bit-rate BR of LoRa becomes:

$$BR = R_M \cdot \frac{4}{4 + CR} = SF \cdot \frac{BW}{2^{SF}} \cdot \frac{4}{4 + CR} \quad bits/s.$$

The more interference bursts are expected, the higher the coding rate that should be used to maximize the probability of successful packet reception. Note that LoRa radios with different coding rates can still communicate, since the packet header (transmitted using the maximum coding rate of 4/8) can include the code rate used for the payload.

Transmission Power (TP). As most wireless radios, LoRa transceivers also allow for adjusting the transmission power, drastically changing the energy required to transmit a packet. By switching the transmission power, for example, from −4 to +20 dBm, the power consumption increases from 66 mW to 396 mW when using the RFM95 transceiver (HopeRF, Shenzhen, China) [26]. Note also that, for transmission powers higher than +17 dBm, hardware limitations and legal regulations limit the radio duty cycle to a maximum of 1%.

Carrier Frequency (CF). LoRa transceivers use sub-GHz frequencies for their communication: among others, the 433 MHz, 868 MHz (Europe), and 915 MHz (North America) industrial, scientific and medical (ISM) radio bands. Common LoRa modules such as the Semtech SX1272 [27] and HopeRF RFM95 [26] support communication in the frequency range [860–1020] MHz and are programmable in steps of 61 Hz. Ten channels with different bandwidths can be used to communicate using LoRa in the European 868 MHz ISM band.

3. Related Work

We now summarize the body of works characterizing the performance of LoRa communications and the effects of environmental conditions on its operations.

Characterization of LoRa performance. Because LoRa technology is closed-source, only a few details about its operations are actually available—mostly derived from Semtech's patent describing the modulation technology or from application notes written to help application designers fine-tune the performance of the transceiver to their needs. Many researchers found this information too limited and started benchmarking and reverse-engineering [28,29] the technology to better understand its mechanism and characteristics.

The first experiments focused on the range of reliable links and on the receiver sensitivity [14]—LoRa's core characteristics. In [11], LoRa has been evaluated in urban and maritime scenarios, and a signal attenuation model was derived. In [5], instead, experiments focused on testing LoRa's communication range on a set of diverse scenarios (from underground to overground, with and without line of sight) in order to provide a set of deployment guidelines.

Interestingly, in the evaluation process, different studies found that results were contradicting Semtech's claims on LoRa performance. In [13], researchers were not able to observe an improved sensitivity with increasing spreading factors. Bor et al. [14] found that LoRa's ability of penetrating buildings is rather limited compared to what was originally claimed. Similarly, the results that we present in this work show that communication in indoor scenarios with no line of sight are among the most challenging conditions for LoRa. Another challenge is represented by vegetation, as found by Iova et al. [12]. Finally, in [14,30,31], the authors model LoRa self-interference and channel utilization, concluding that LoRa's scalability is worse than what was originally promised. Other works focus on a more detailed characterization of LoRa, in particular on packet loss [32], on the ability of receiving packets from concurrent transmissions [7], and on the energy consumption at different transmission powers [4].

Different from previous works, this paper analyzes LoRa's PHY settings from a multi-objective perspective, with the goal of finding the best trade-off between data rate, packet reception rate, and energy efficiency.

Environmental effects on low-power radios. A large body of works has studied the impact of environmental conditions on network performance in low-power wireless radios, especially on IEEE 802.15.4-compliant radio transceivers. Several authors report the impact of meteorological conditions on packet reception, including the impact of weather conditions [20,33,34], and humidity [35], as well as the presence of vegetation [36]. One of the most comprehensive studies on wireless nodes deployed outdoors was carried out by Wennerström et al. [20], who have highlighted that packet reception ratio and received signal strength correlate the most with temperature, whereas the correlation with other factors such as absolute humidity and precipitation is less pronounced.

The strong impact of temperature on communication performance has been confirmed by several other works, also almost entirely focused on IEEE 802.15.4 transceivers. Bannister et al. [19] have shown the correlation between temperature and signal strength in a deployment in the Sonoran desert, and identified in a temperature-controlled chamber that the received signal strength of the TI CC2420 radio attenuates at high temperatures due to the impact of temperature on the radio's low-noise and power amplifiers. Based upon this work, Boano et al. [17,18] have confirmed these findings also on other platforms such as the TI CC1020 and CC2520, and also highlighted how this can cause a complete disruption of a wireless link. The authors have also shown how the impact of temperature cannot be neglected when designing duty-cycled medium access control protocols for low-power wireless radios [37,38]. To facilitate the study of how temperature affects the operation of low-power wireless protocols on a larger scale than in a temperature-controlled chamber, several low-cost testbed infrastructures have been proposed, the most popular being TempLab and HotBox [39,40].

The impact of environmental conditions on LPWAN radios, instead, has not yet been investigated in detail. Iova et al. [12] have deployed a number of LoRa networks in urban and mountain environments, and reported that environmental factors such as the presence of vegetation and temperature variations can negatively affect communication performance. The authors, however, did not quantify the impact of these environmental factors and their work does not yet clarify whether high temperatures degrade the quality of LoRa links in a similar way as observed on several IEEE 802.15.4 transceiver platforms.

In the remainder of this paper, we conduct a number of experiments to complement the body of aforementioned related works and answer two key questions that are yet open: (i) how does the selection of LoRa's PHY settings affect the efficiency of links, including the ones of intermediate quality? and (ii) how does temperature affect the performance of LoRa? To answer these questions, we start by carrying out experiments indoor, outdoor, and underground, and by analyzing how PHY settings and environmental factors affect LoRa's communication performance.

4. Evaluating the Performance of LoRa

To study the reliability and energy-efficiency of LoRa communications as a function of the PHY settings described in Section 2 and as a function of environmental factors, we conduct a series of small-scale deployments.

Experimental setup. All of our experiments are carried out at the Graz University of Technology, Austria: the exact location of the nodes is shown in Figure 1. We fix the senders at three given locations (S) and place three receivers at different distances (1, 2, 3) for three scenarios: *indoor* with obstacles (i), *outdoor* with direct line of sight (o), and *underground* covered by a metal manhole (u). Each transmitter sends a packet with a 5-byte payload every 3 s at transmission power +20 dBm, emulating a timely report of a typical IoT sensor for urban monitoring. Every six minutes, transmitter and receivers reboot and switch to a different setting according to a set of hard-coded combinations shown in Table 2. For each of the three scenarios (indoor, outdoor, and underground), we test each setting configuration sequentially every six minutes for a duration of 24 h, hence resulting in a total of 1600 packets exchanged per setting.

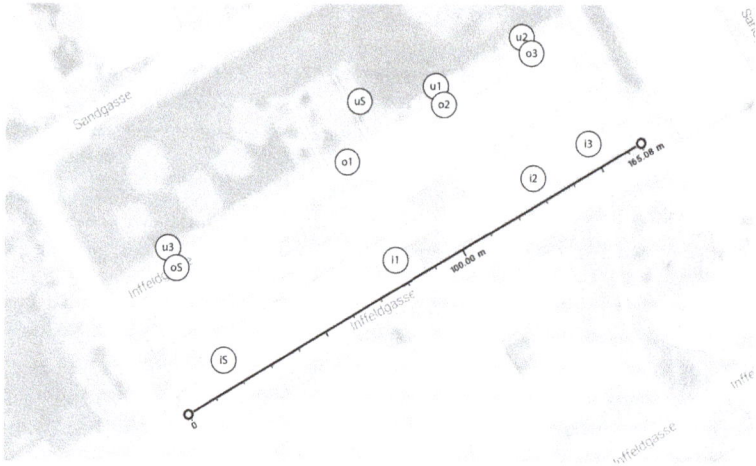

Figure 1. Deployment map for our experiments indoor (**i**), outdoor (**o**), and underground (**u**). The sender node for each scenario is indicated with iS, oS, and uS, respectively.

Table 2. LoRa settings used in our experiments: spreading factor (SF), code rate (CR), bandwidth (BW), and rit-rate (BR) Note that settings are ordered by decreasing bit-rate.

Setting ID	1	2	3	4	5	6	7	8	9	10	11	12	13	14	15	16	17	18
SF	7	7	7	9	7	7	9	9	7	9	9	12	9	12	12	12	12	12
CR	4/5	4/8	4/5	4/5	4/8	4/5	4/8	4/5	4/8	4/8	4/5	4/5	4/8	4/8	4/5	4/8	4/5	4/8
BW (kHz)	500	500	250	500	250	125	500	250	125	125	125	500	125	500	250	250	125	125
BR (kb/s)	21.87	13.62	10.93	7.03	6.83	5.47	4.39	3.51	3.41	2.2	1.76	1.16	1.09	0.72	0.58	0.37	0.30	0.18

Figure 2. Custom-built LoRa platform based on the Moteino MEGA (LowPowerLab, Canton, MI, USA) inside a water-proof enclosure (top removed) [41].

Hardware. The experiments are conducted using a custom-built platform (see Figure 2) based on the Moteino MEGA (LowPowerLab, Canton, MI, USA) [42]. The latter is equipped with an ATMega1284P microcontroller, and a HopeRF RFM95 LoRa transceiver operating at 868 MHz [26]. The device is powered by a 3.7 V Li-Ion battery with a capacity of 3.4 Ah that can be charged via a dedicated circuit. Without duty cycling the radio, this battery can sustain the device operation for more than 24 h, the maximum duration of our experiments. The platform we have built also embeds sensors to measure changes in the surrounding environment. In particular, temperature and humidity are read from a Bosch BME280 sensor (Gerlingen, Germany) via the I2C interface. For persistent storage, an SD card logs each received packet together with its sequence number, the sensed environmental conditions, as well as the time-stamp provided by a Maxim DS3231 real-time clock (San Jose, CA,

USA). We also save the presence of cyclic redundancy check (CRC) errors in the received packets in our traces. This hardware setup was used in our experiments both for senders and receivers.

Metrics. For each 6-min experiment, we compute the packet reception ratio (*prr*) and the receiver sensitivity, i.e., the lowest signal strength among successfully received packets. We then check the correlation of the computed *prr* with the employed PHY settings, as well as with the measured temperature, humidity, and received signal strength values.

4.1. Reliability of LoRa as a Function of PHY Settings

Figure 3 shows the packet reception ratio (i.e., the percentage of packets sent that were correctly received) indoor, outdoor and underground for a number of different radio settings (see Table 2). Figure 3 plots all 6-min experiments grouped by setting ID.

Horizontal red lines represent the median, while blue boxes represent the 25th and 75th percentiles. The remaining results are enclosed by vertical dashed black lines while statistical outliers are represented by red crosses. Note that these results were previously presented in [41].

Our results show that LoRa setting ID 11 (i.e., BW = 125, SF = 9, and CR = 4/5) achieves a packet reception ratio above 95% regardless of the scenario and distance between nodes. Nevertheless, setting ID 2 also performs remarkably well: although it sustains a lower *prr*, it sends packets using a bit-rate that is almost eight times faster than the one used by setting ID 11. This observation will be the starting point of our analysis in Section 5 answering the question of whether it is worth selecting PHY settings that reduce the data rate in order to maximize the link quality.

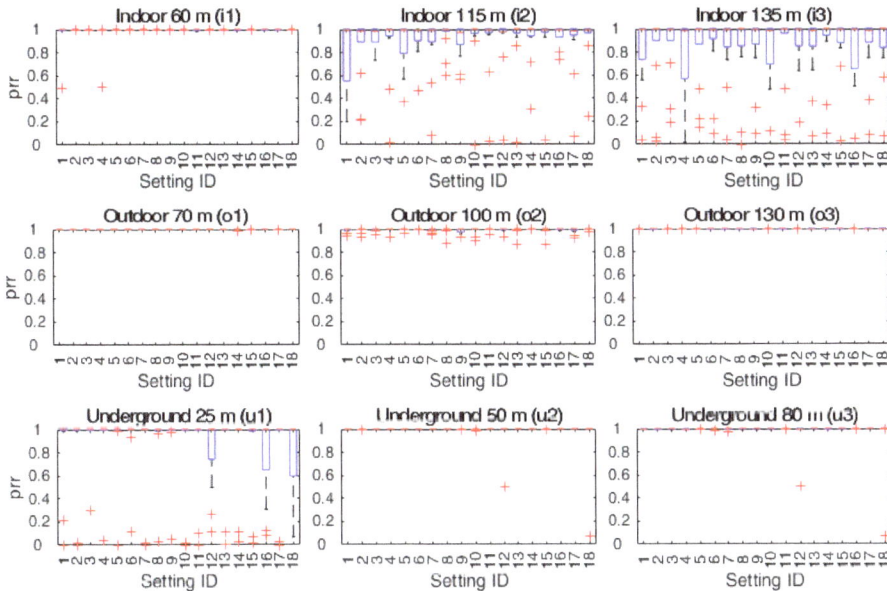

Figure 3. Packet reception ratio *prr*, i.e. the fraction of transmitted packets that are successfully received, for different distances, scenarios, and physical (PHY) settings (setting ID). Horizontal red lines represent the median, while blue boxes represent the 25th and 75th percentiles. The remaining results are enclosed by vertical dashed black lines while statistical outliers are represented by red crosses. Each node is positioned according to Figure 1.

Differently from previous works, in Figure 3, we also show the distribution of the experiment results rather than just the median and mean values. This allows us to make three observations. First, while the median *prr* is close to 1 for most settings, the quartiles and minima are not. Second, due to the lower multi-path and fading effects outdoor and underground, LoRa communications are more reliable in these scenarios rather than indoors (in line with what is observed in [7]), with packet reception ratios above 97% for almost all setting IDs. Third, the range of the LoRa radios is consistent throughout the different settings: even though the reception rate changes, all settings are able to deliver packets in similar conditions.

4.2. Factors Affecting LoRa Reliability

We explore next which environmental factors affect the reliability of LoRa. Towards this goal, we use the traces collected in the previous experiments and focus on the correlation between the packet reception rate (*prr*), setting ID (*set*), temperature (*temp*), humidity (*hum*), spreading factor (*sf*), coding rate (*cr*), bandwidth (*bw*), receiver sensitivity (*sens*), receiver signal strength (*rss*), and hour of the day (*hour*). In particular, we plot the Pearson correlation of each pair of parameters for different experimental scenarios in Figure 4: a value close to 1 (black) means that the two parameters are linearly correlated, whereas a value of 0 (white) implies that the two parameters are independent.

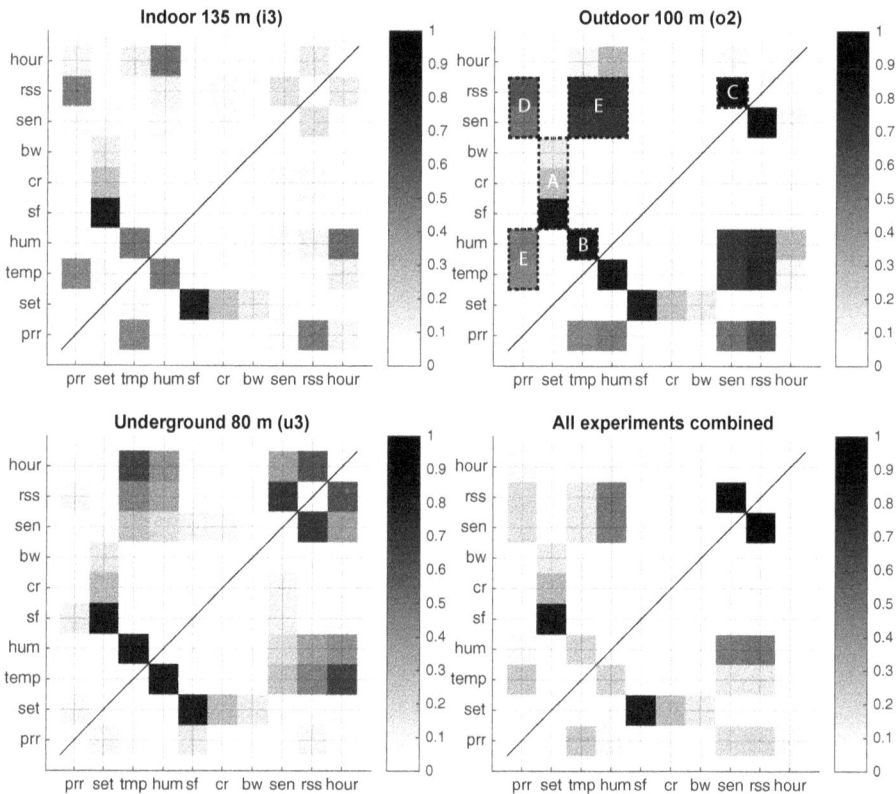

Figure 4. Correlation matrix for different LoRa settings indoor, outdoor, and underground. The plot on the bottom right combines all nine of the settings from Figure 3.

Figure 4 shows that, in all three scenarios (indoor, outdoor and underground), there are some obvious correlations. First (**A**), the setting ID depends on the bandwidth, coding rate (*cr*), and spreading factor (*sf*). This is to be expected, because the setting ID unequivocally describes a combination of these three PHY parameters. Second (**B**), temperature (*temp*) is highly correlated with humidity (*hum*) and both are correlated with the time of the day (*hour*). This is also an expected correlation, as these environmental factors are highly dependent on the sun exposure. Third (**C**), the radio sensitivity (*sen*) is correlated to the received signal strength (*rss*), since the former is defined as the minimum of the latter. Furthermore, one can also note in Figure 4 that the received signal strength (*rss*) is correlated with the packet reception ratio (*prr*) (**D**), as the LoRa radio is able to successfully decode packets that are above a certain signal-to-noise ratio.

Figure 4 also shows that temperature is tightly correlated with the received signal strength (*rss*) and the packet reception ratio (*prr*). This seem to hint that temperature variations may affect the operation of the employed LoRa radio in a similar way as observed on some IEEE 802.15.4 transceivers (see Section 3). When analyzing the figure in detail, one can actually observe a correlation cluster (**E**) between temperature, humidity, time of the day, packet reception ratio, and received signal strength: the strength of these correlations varies depending on the scenario and is stronger outdoors. To better understand the inter-dependency between the reliability performance (*rss* and *prr*) and environmental factors (temp), we carry out experiments in controlled settings in Section 6.

5. The Efficiency of LoRa as a Function of PHY Settings

The experimental campaign presented in the previous section has shown that it is possible to improve the reliability of LoRa by carefully choosing the PHY settings, i.e., some of the settings allow for sustaining a higher *prr*. In this section, we analyze the costs of such improvement in terms of energy efficiency and analyze in detail the trade-off between packet delivery rate and setting's bandwidth, providing an answer to the question: *is it more efficient to use resilient and slow settings or to use faster (but more fragile) configurations together with a re-transmission mechanism?*

To answer this question, we focus on the most challenging scenario in our experimental campaign, i.e., indoor, no line of sight, and with a distance between two devices of 115 m. Figure 5a shows the distribution of packet reception ratios as a function of setting ID. Averages are represented by '*', while median, quartiles, and extreme values are enclosed by a blue box and two black bars (outliers are indicated with crosses). PHY settings are ordered by decreasing bit-rate, from faster and more lightweight settings on the left, to slower settings increasing the transceiver's on-time on the right. As we can see, using the fastest setting (setting ID 1 and 2), the average *prr* is 80% with a worst case scenario where *prr* is as low as 20%. As expected, by selecting a PHY configuration that reduces the bit-rate (i.e., by decreasing the bandwidth and increasing the bit redundancy), the packet reception ratio improves, as well as its distribution.

Nevertheless, one can argue that it is more energy-efficient to re-transmit a packet using the faster settings available rather than employing PHY settings that reduce the bit-rate. To prove our point, we compute the expected number of re transmissions (*ETX*) as:

$$ETX = \frac{1}{prr}$$

and compare them against each setting's original bit-rate (*BR*). As we can see in Figure 5b, the expected number of re-transmissions (squares) does not directly depend on the settings' bit-rate (triangles), suggesting that not all settings are worth their overhead.

In order to give an indication on how efficiently LoRa settings trade communication efficiency against reliability, we compute the effective bit-rate (*EBR*) as:

$$EBR[kb/s] = \frac{BR}{ETX} = BR \cdot prr$$

and show both mean and distribution in Figure 5c.

The EBR shows the expected bit-rate of each setting in the case packets are re-transmitted back-to-back until one is successfully received. As we can see from Figure 5c, the mean EBR is by far the highest when the fastest LoRa setting is used (setting ID 1). Setting ID 2, on the other hand, is more consistent, showing a lower variance and the highest minimum (crosses represent outliers).

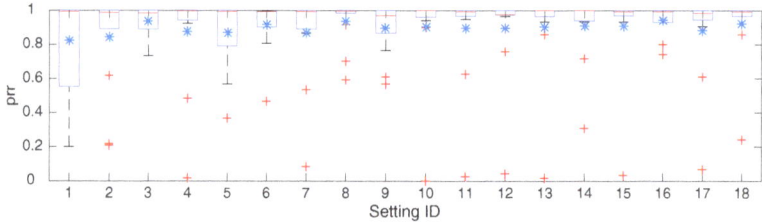

(a) Packet reception rate *(prr)* for different LoRa settings. The higher the *(prr)*, the better.

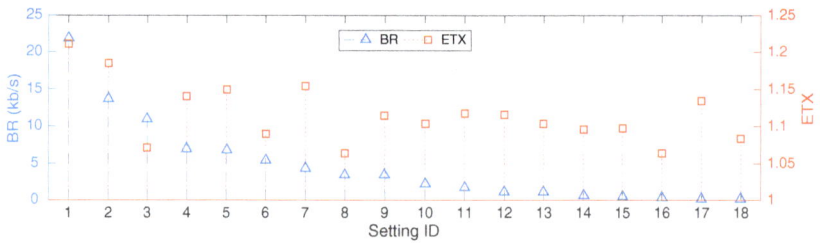

(b) Bit-rate (BR) versus expected number of retransmissions (ETX) computed as $\frac{1}{prr}$.

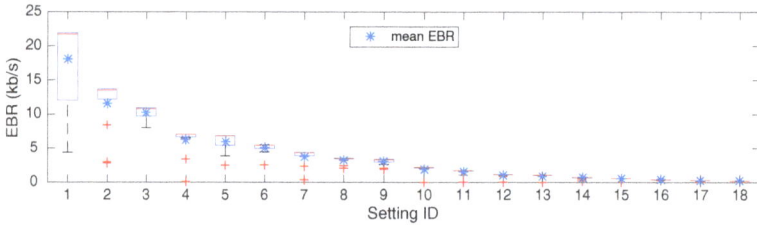

(c) Expected bit-rate computed as $\frac{BR}{ETX}$.

Figure 5. LoRa performance as a function of PHY settings in an indoor scenario without line of sight at a distance of 115 m. For (**a**,**c**), asterisks represent mean values, horizontal red lines represent the median, while blue boxes represent the 25th and 75th percentiles. The remaining results are enclosed by vertical dashed black lines while statistical outliers are represented by red crosses.

Even though these results can heavily depend on the surrounding environment, we argue that, *in order to maximize the effective bit-rate, one should opt for a re-transmission mechanism and use the settings with sufficient prr (e.g., >0.2) and the highest bit-rate possible.*

To test the validity of this claim, we additionally compute the EBR for 12 experiments run by independent researchers [5] and present the results in Figure 6. The experiment was run on several Libelium Waspmote LoRa motes; therefore, the setting ID shown in the figure (mode ID) is enumerated according to the Libelium application programming interface (API). As for Figure 5, we conveniently order the PHY settings starting from the one using the highest bit-rate on the left, to the one employing the lowest bit-rate on the right. As we can see, this set of experiments also confirms our observation: *faster settings result in higher bit-rates, even though the quality of the link (i.e., the (prr)) is lower.*

Next, we extend this analysis to different transmission power levels and explicitly also evaluate the energy efficiency of LoRa transmissions.

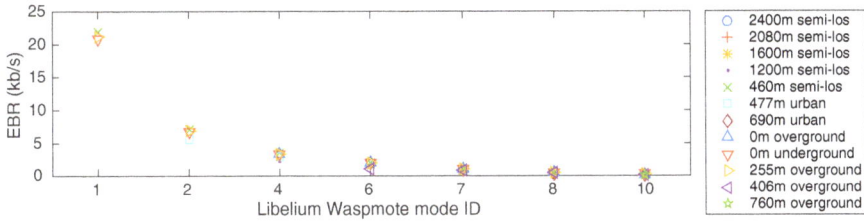

Figure 6. Effective bit-rate for 12 experiments run by independent researchers and presented in [5].

Energy efficiency of LoRa transmissions. We evaluate the energy efficiency of different PHY settings by repeating the indoor experiments without line of sight at a distance of 115 m. This time, we slightly vary the experimental setup as follows: we position the two nodes at the edge of transmission range and send each packet with a different power level, sequentially selected from the set {+20, +17, +15, +11, +7, +5} dBm. We first check if using the fastest PHY setting leads to the highest energy efficiency both in challenging and non challenging conditions (the lower the transmission power, the more challenging the communication). We then compute the most energy-efficient setting by computing the effective energy consumed to send a kilobit of data *EKB* as follows:

$$EKB[J/kb] = \frac{P}{EBR},$$

where *P* is the power consumption of the radio in watts and EBR is the effective bit rate.

Figure 7a shows the packet reception rate when using three different power levels: +20, +15, and +11 dBm. As expected, changing the transmission power drastically affects the *prr*, since the nodes are intentionally placed on the edge of the communication range. In agreement with the results presented previously, Figure 7b shows that the fastest settings are the ones with highest effective bit-rate *EBR*—*independently of the employed transmission power*. As lower transmission powers imply lower energy expenditures, we still need to answer *which transmission power configuration results in the highest energy efficiency*.

Figure 7c shows the energy required by each PHY setting to transmit a kilobit of data (*EKB*), including the cost of the re-transmissions. As we can see, the most efficient transmission power configuration (i.e., leading to the lowest EKB) is the highest, i.e., +20 dBm. Therefore, our experimental results suggest that, *together with the fastest setting, the highest transmission power should be preferred*: this combination provides the highest bit-rate *EBR* and the lowest energy consumption *EKB*. It is worth highlighting that, in less challenging scenarios in which several transmission powers achieve a *prr* = 1, the lowest one should be used, as the higher transmission powers may increase the energy consumption without any additional benefit.

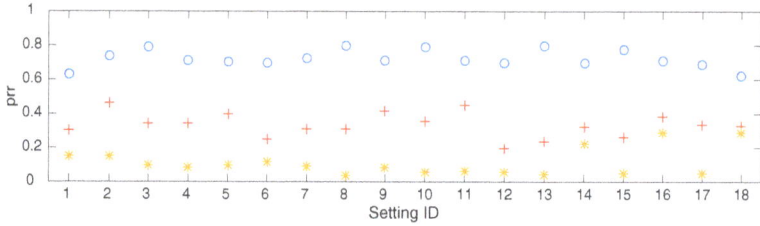

(a) Packet reception rate ((prr)) for different LoRa settings. The higher the (prr), the better.

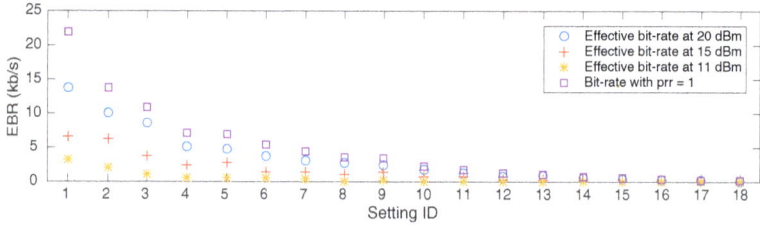

(b) Expected bit-rate computed as $\frac{BR}{ETX}$.

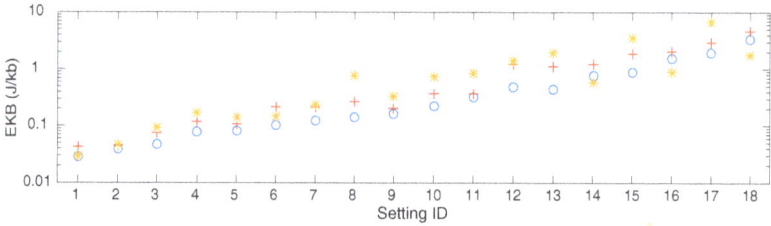

(c) Expected energy per kilobit (EKB). The lower EKB, the more energy efficient is LoRa.

Figure 7. LoRa performance for different settings and transmission powers.

6. Impact of Temperature on LoRa Transceivers

Our experimental campaign presented in Section 4 has shown that there is a strong correlation between temperature, packet reception rate, and received signal strength. To quantify this correlation and shed light on the impact of temperature on LoRa communications, we carry out a deeper investigation in controlled settings.

Experimental setup. We use the TempLab testbed [39] to expose a number of LoRa nodes to repeatable temperature variations as shown in Figure 8. The TempLab testbed available at Graz University of Technology has two different types of nodes [38]: LO nodes only heating the sensor nodes above room temperature and PE nodes having the capability to also cool down the node's temperature below zero degrees thanks to enclosures made of hard Polystyrene foam and ATA-050-24 Peltier air-to-air assembly modules (Custom Thermoelectric, Bishopville, MD, USA). Both LO and PE nodes are remotely controlled using an Aeon Z-Wave Stick Series 2 (Aeon Labs LLC, El Cerrito, CA, USA) sending commands to (i) Vesternet EVR_AD1422 Z-Wave Everspring wireless dimmers (Smartech Holdings Ltd., Manchester, UK) connected to Philips E27 infra-red 100 W light bulbs (Philips, Eindhoven, The Netherlands) and (ii) to Vesternet EVR_AN1572 Z-Wave Everspring on–off wireless switches connected to the Peltier modules.

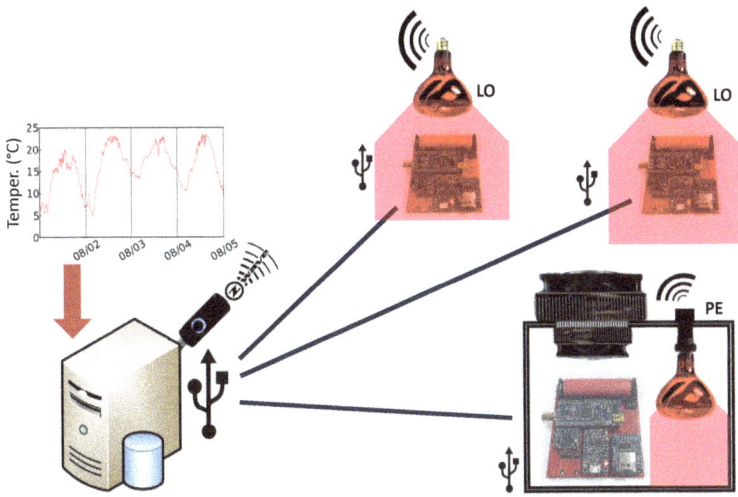

Figure 8. Sketch of the employed TempLab's setup to control the temperature of LoRa nodes.

We place the LoRa nodes without a connected SubMiniature version A (SMA) antenna inside PE nodes and let the nodes transmit packets as fast as possible without any radio duty cycling while temperature varies in the range of 0–60 °C. In particular, we scripted TempLab to first slowly increase temperature from 0 to 60 °C and then to quickly cool down to 0 °C. Each test has a duration of five hours and was repeated for different PHY settings and hardware platforms.

Impact on received signal strength. We plot the relationship between the received signal strength indicator (RSSI) of received packets and the median temperature measured by the two nodes for different hardware platforms. Figure 9 shows the curve recorded when using a Moteino MEGA board [42] equipped with a HopeRF RFM95 LoRa radio (the same platform used for the experiments described in Section 4). Each dot represents the median of the RSSI over 40 received packets. Similarly to what was reported in [18], the RSSI decreases linearly in discrete steps for a total of about 6 dB in the temperature range of 0–60 °C. This is because, for a given voltage, a higher temperature increases the resistance of conductors, while reducing the pass-trough current. For radio transceivers, this implies that higher temperatures reduce the received signal strength and signal-to-noise ratio.

Also according to [18], we observe an hysteresis in the relationship between RSSI and temperature when comparing the curves obtained when heating and when cooling the LoRa nodes. Similarly, Figure 10 shows the relationship between RSSI and temperature recorded when exposing ST Nucleo L073RZ boards (STMicroelectronics, Geneva, Switzerland) [9] equipped with a Semtech SX1272 radio to temperature variations. Also using this hardware, we observe a linear decrease of about 6 dB in the RSSI at high temperatures. We attribute the spikes recorded on the experiment of Figure 10 (when temperature varies between 30 and 40 °C) to a temporary multi-path fading effect of the environment. Note that Figures 9 and 10 refer to different experiments, both carried out using setting ID 6 (i.e., CR = 4/5, SF = 7, BW = 125 MHz).

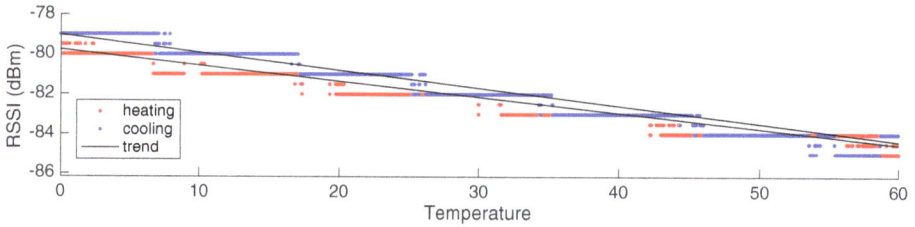

Figure 9. Received signal strength indicator (RSSI) as a function of temperature on the Moteino MEGA platform employing a HopeRF RFM95 transceiver [42].

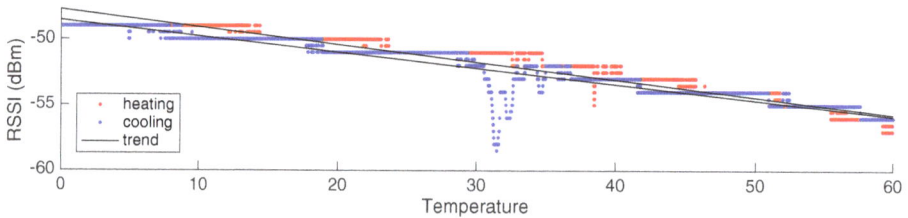

Figure 10. RSSI as a function of temperature on the ST Microelectronics Nucleo L073RZ platform employing a Semtech SX1272 transceiver [9].

Remarkably, the attenuation of received signal strength caused by an increase of temperature in the range [0–60] °C is comparable to the change in sensitivity that can be observed when switching from the fastest to the slowest PHY setting [14]. Therefore, in case of cold temperatures, it may even be possible to avoid using extremely slow radio settings by carefully deploying the LoRa devices in locations that are not directly exposed to sunlight.

Impact on packet reception ratio. We further analyze the effects of temperature variations on nodes that are at the edge of their communication range. We intentionally place two nodes at the limits of their communicating range and slowly change the temperature of the transmitter from 15 to 60 °C and quickly back to 15 °C. Figure 11 shows the distribution of lost, corrupted and successfully received packets for every minute in a 75-min experiment. We can observe that what was a perfect link at minute 0 (100% *prr* at 15 °C) slowly becomes unusable at higher temperatures. As soon as temperature (red line) starts to increase, either packets are received, but their content is corrupted, or the radio was unable to receive the packet at all. Once temperature starts decreasing again, the link is restored and sustains a high delivery rate. These experiments confirm results from previous studies on specific IEEE 802.15.4 radios [18,19], and show that temperature can drastically affect packet delivery. An important takeaway message is that LoRa nodes employing the radio transceivers used in our experiments should be deployed during the warmest time of the day or year, to ensure that network performance is sufficient throughout the system lifetime, and that nodes should be shielded from sunlight if possible.

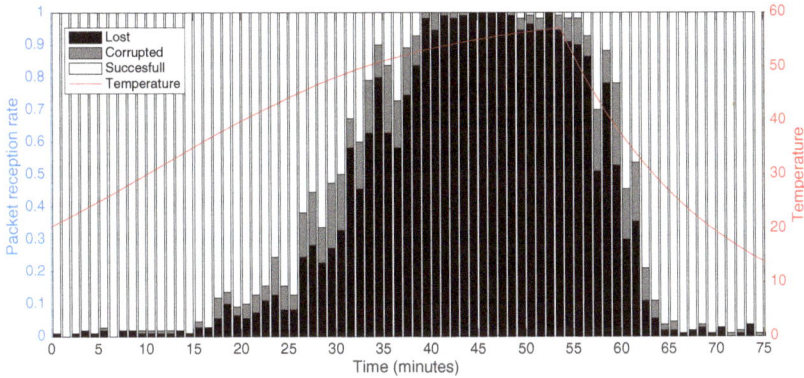

Figure 11. Increase of packet corruption and loss at higher temperature on a LoRa link at the edge of the communication range.

7. Conclusions

This paper presents an analysis of the performance of LoRa as a function of different PHY settings and environmental conditions. We first study the effects of different LoRa settings on the effective bit-rate that can be achieved (i.e., on the amount of information that LoRa is able to *successfully* deliver during a given period). Our experimental results suggest that, when nodes are at the edge of their communication range, using the fastest PHY setting and the highest transmission power is more efficient than selecting slower settings that maximize the link quality. Even though, for example, the fastest PHY setting in our experiments yields an *average* packet reception rate that is 10% lower than the slowest setting, the former's *effective bitrate* is 100× faster than the latter's. Compared to the slower settings, the efficiency of the fastest PHY setting is so high that even in its worst case scenario—when the *minimum prr* reaches 20%—the effective bitrate is faster than twelve of the slowest PHY settings (settings 7–18, from 1× to 25× better). Second, we analyze the external factors affecting the reliability of LoRa. Our outdoor experiments show a clear correlation between temperature, humidity, packet reception rate, and received signal strength. A deeper investigation in controlled settings shows that the signal strength of received packets decreases linearly when temperature increases in two different LoRa transceivers. Different LoRa radios have shown that, over a range of 60 °C, the received signal strength is consistently reduced by 6 dBm (1 dBm/10 °C). This decrease in signal strength can significantly affect LoRa links that are at the edge of the communication range, increasing packet corruption and loss, and rendering a perfectly good link (100% *prr* at 15 °C) completely unusable (0% *prr* at 60 °C).

As a future work, we plan to quantify the impact of other environmental factors on LoRa performance, e.g., humidity and radio interference. Our ultimate goal is to design and implement an environmental-aware MAC protocol tailored to LoRa that can sustain reliable and energy-efficient operations regardless of changes in the surrounding environmental conditions.

Acknowledgments: This work has been supported by the Sino Austrian Electronic Technology Innovation Center. This work was also partially performed within the LEAD-Project "Dependable Internet of Things in Adverse Environments", funded by Graz University of Technology.

Author Contributions: Marco Cattani has conceived and designed the experiments evaluating the interplay between LoRa's PHY settings and link quality described in Sections 4 and 5, as well as taken care of the write-up of the whole manuscript. Carlo Alberto Boano has conceived and designed the experiments evaluating the impact of temperature on LoRa transceivers described in Section 6 and taken care of the write-up of the whole manuscript. Kay Römer has participated to the general discussions and help revising the write-up of the manuscript.

Conflicts of Interest: The authors declare no conflict of interest.

References

1. Real Wireless Ltd. *A Comparison of UNB and Spread Spectrum Wireless Technologies as Used in LPWA M2M Applications*; Real Wireless: West Sussex, UK, 2015.
2. LoRa Alliance. LoRa: Wide Area Networks for IoT, 2017. Available online: http://www.lora-alliance.org/What-Is-LoRa/Technology (accessed on 22 May 2017).
3. Raza, U.; Kulkarni, P.; Sooriyabandara, M. Low Power Wide Area Networks: An Overview. *IEEE Commun. Surv. Tutor.* **2017**, *19*, 855–873.
4. Bor, M.; Roedig, U. LoRa Transmission Parameter Selection. In Proceedings of the 13th IEEE International Conference on Distributed Computing in Sensor Systems (DCOSS), Ottawa, ON, Canada, 5–7 June 2017.
5. Gnawali, O.; Fonseca, R.; Jamieson, K.; Moss, D.; Levis, P. Demystifying Low-Power Wide-Area Communications for City IoT Applications. In Proceedings of the 10th ACM Workshop on Wireless Network Testbeds, Experimental Evaluation, and Characterization (WiNTECH), New York, NY, USA, 3 October 2016; pp. 2–8.
6. Baños-Gonzalez, V.; Afaqui, M.S.; López-Aguilera, E.; Villegas, E.G. Throughput and Range Characterization of IEEE 802.11ah. *arXiv* **2016**, arXiv:1604.08625.
7. Bor, M.; Vidler, J.; Roedig, U. LoRa for the Internet of Things. In Proceedings of the 1st International Workshop on New Wireless Communication Paradigms for the Internet of Things (MadCom), Graz, Austria, 15–17 February 2016; pp. 361–366.
8. Libelium. *Waspmote-LoRa-868MHz-915MHz-SX1272 Networking Guide, v7.0*; Libelium: Zaragoza, Spain, 2017.
9. ST Microelectronics. *STM32 Nucleo Pack for LoRa Technology (P-NUCLEO-LRWAN1), DocID029505 Rev. 2*; ST Microelectronics: Geneva, Switzerland, 2016.
10. NetBlocks Embedded Networking. XRange SX1272 LoRa RF module. Available online: http://www.netblocks.eu/ (accessed on 22 May 2017).
11. Petäjäjärvi, J.; Mikhaylov, K.; Roivainen, A.; Hänninen, T.; Pettissalo, M. On the Coverage of LPWANs: Range Evaluation and Channel Attenuation Model for LoRa Technology. In Proceedings of the 14th IEEE International Conference on ITS Telecommunications (ITST), Copenhagen, Denmark, 2–4 December 2015; pp. 55–59.
12. Iova, O.; Murphy, A.L.; Ghiro, L.; Molteni, D.; Ossi, F.; Cagnacci, F. LoRa from the City to the Mountains: Exploration of Hardware and Environmental Factors. In Proceedings of the 2nd International Workshop on New Wireless Communication Paradigms for the Internet of Things (MadCom), Uppsala, Sweden, 20–22 February 2017.
13. Augustin, A.; Yi, J.; Clausen, T.; Townsley, W.M. A Study of LoRa: Long Range & Low Power Networks for the Internet of Things. *Sensors* **2016**, *16*, 1466.
14. Bor, M.; Roedig, U.; Voigt, T.; Alonso, J.M. Do LoRa Low-Power Wide-Area Networks Scale? In Proceedings of the 19th ACM International Conference on Modeling, Analysis and Simulation of Wireless and Mobile Systems (MSWiM), Valletta, Malta, 13–17 November 2016; pp. 59–67.
15. Baccour, N.; Koubâa, A.; Mottola, L.; Youssef, H.; Zúñiga, M.A.; Boano, C.A.; Alves, M. Radio Link Quality Estimation in Wireless Sensor Networks: A Survey. *ACM Trans. Sensor Netw.* **2012**, *8*, 34.
16. Gnawali, O.; Fonseca, R.; Jamieson, K.; Moss, D.; Levis, P. Collection Tree Protocol. In Proceedings of the 7th International Conference on Embedded Networked Sensor Systems (SenSys), Berkeley, CA, USA, 4–6 November 2009; pp. 1–14.
17. Boano, C.A.; Brown, J.; Tsiftes, N.; Roedig, U.; Voigt, T. The Impact of Temperature on Outdoor Industrial Sensornet Applications. *IEEE Trans. Ind. Inform.* **2010**, *6*, 451–459.
18. Boano, C.A.; Wennerström, H.; Zúñiga, M.A.; Brown, J.; Keppitiyagama, C.; Oppermann, F.J.; Roedig, U.; Nordén, L.Å.; Voigt, T.; Römer, K. Hot Packets: A Systematic Evaluation of the Effect of Temperature on Low Power Wireless Transceivers. In Proceedings of the 5th Extreme Conference on Communication (ExtremeCom), Reykjavik, Iceland, 24–29 August 2013; pp. 7–12.
19. Bannister, K.; Giorgetti, G.; Gupta, S.K. Wireless Sensor Networking for Hot Applications: Effects of Temperature on Signal Strength, Data Collection and Localization. In Proceedings of the 5th International Workshop on Embedded Networked Sensors (HotEmNets), Charlottesville, VA, USA, 2–3 June 2008.

20. Wennerström, H.; Hermans, F.; Rensfelt, O.; Rohner, C.; Nordén, L.A. A Long-Term Study of Correlations between Meteorological Conditions and 802.15.4 Link Performance. In Proceedings of the 10th IEEE International Conference on Sensing, Communication, and Networking (SECON), New Orleans, LA, USA, 24–27 June 2013; pp. 221–229.

21. Ratasuk, R.; Vejlgaard, B.; Mangalvedhe, N.; Ghosh, A. NB-IoT system for M2M communication. In Proceedings of the IEEE Wireless Communications and Networking Conference (WCNC), Doha, Qatar, 3–6 April 2016; pp. 1–5.

22. Weightless SIG. Weightless-P Open Standard, 2017. Available online: http://www.weightless.org/about/weightlessp (accessed on 22 May 2017).

23. Sigfox. Sigfox Technology, 2017. Available online: http://www.sigfox.com (accessed on 22 May 2017).

24. Weightless SIG. Weightless-N Open Standard, 2017. Available online: http://www.weightless.org/about/weightlessn (accessed on 22 May 2017).

25. Semtech Corporation. *LoRa Modulation Basics—Application Note 1200.22, Revision 2*; Semtech Corporation: Camarillo, CA, USA, 2015.

26. Hope RF Microelectronics. *RFM95/96/97/98(W)—Low Power Long Range Transceiver Module, v1.0*; Hope RF Microelectronics: Shenzhen, China, 2016.

27. Semtech Corporation. *SX1272/73—860 MHz to 1020 MHz Low-Power Long-Range Transceiver, Revision 3.1*; Semtech Corporation: Camarillo, CA, USA, 2017.

28. Myriad-RF. LoRa-SDR, 2017. Available online: http://github.com/myriadrf/LoRa-SDR (accessed on 22 May 2017).

29. DecodingLora, 2017. Available online: http://revspace.nl/DecodingLora (accessed on 22 May 2017).

30. Georgiou, O.; Raza, U. Low Power Wide Area Network Analysis: Can LoRa Scale? *IEEE Wirel. Commun. Lett.* **2017**, *6*, 162–165.

31. Voigt, T.; Bor, M.; Roedig, U.; Alonso, J. Mitigating Inter-network Interference in LoRa Networks. In Proceedings of the 2nd International Workshop on New Wireless Communication Paradigms for the Internet of Things (MadCom), Uppsala, Sweden, 20–22 February 2017.

32. Marcelis, P.; Rao, V.; Prasad, R.V. DaRe: Data Recovery through Application Layer Coding for LoRaWAN. In Proceedings of the 2nd International Conference on Internet-of-Things Design and Implementation (IoTDI), Pittsburgh, PA, USA, 18–21 April 2017; pp. 97–108.

33. Anastasi, G.; Falchi, A.; Passarella, A.; Conti, M.; Gregori, E. Performance Measurements of Motes Sensor Networks. In Proceedings of the 7th ACM International Symposium on Modeling, Analysis and Simulation of Wireless and Mobile Systems (MSWiM), Venice, Italy, 4–6 October 2004; pp. 174–181.

34. Boano, C.A. Application Support Design for Wireless Sensor Networks. Master's Thesis, Politecnico di Torino, Turin, Italy; Kungliga Tekniska Högskolan, Stockholm, Sweden, 2009.

35. Thelen, J.; Goense, D.; Langendoen, K. Radio Wave Propagation in Potato Fields. In Proceedings of the 1st Workshop on Wireless Network Measurement (WiNMee), Garda, Italy, 1–5 April 2005.

36. Marfievici, R.; Murphy, A.L.; Picco, G.P.; Ossi, F.; Cagnacci, F. How Environmental Factors Impact Outdoor Wireless Sensor Networks: A Case Study. In Proceedings of the 10th IEEE International Conference on Mobile Ad-Hoc and Sensor Systems (MASS), Hangzhou, China, 14–16 October 2013; pp. 565–573.

37. Boano, C.A.; Römer, K.; Tsiftes, N. Mitigating the Adverse Effects of Temperature on Low-Power Wireless Protocols. In Proceedings of the 11th IEEE International Conference on Mobile Ad Hoc and Sensor Systems (MASS), Philadelphia, PA, USA, 27–30 October 2014; pp. 336–344.

38. Boano, C.A. Dependable Wireless Sensor Networks. Ph.D. Thesis, Graz University of Technology, Graz, Austria, 2014.

39. Boano, C.A.; Zúñiga, M.A.; Brown, J.; Roedig, U.; Keppitiyagama, C.; Römer, K. TempLab: A Testbed Infrastructure to Study the Impact of Temperature on Wireless Sensor Networks. In Proceedings of the 13th ACM/IEEE International Conference on Information Processing in Sensor Networks (IPSN), Berlin, Germany, 15–17 April 2014; pp. 95–106.

40. Schmidt, F.; Ceriotti, M.; Hauser, N.; Wehrle, K. If You Can't Take the Heat: Temperature Effects on Low-Power Wireless Networks and How to Mitigate Them. In Proceedings of the 12th European Conference on Wireless Sensor Networks (EWSN), Porto, Portugal, 9–11 February 2015.

41. Cattani, M.; Boano, C.A.; Steffelbauer, D.; Kaltenbacher, S.; Günther, M.; Römer, K.; Fuchs-Hanusch, D.; Horn, M. Adige: An Efficient Smart Water Network based on Long-Range Wireless Technology. In Proceedings of the 3rd International Workshop on Cyber-Physical Systems for Smart Water Networks (CySWATER), Pittsburgh, PA, USA, 18–21 April 2017.

42. LowPowerLab. Moteino MEGA LoRa, 2016. Available online: http://lowpowerlab.com/shop/product/119 (accessed on 22 May 2017).

Journal of
*Sensor and
Actuator Networks*

MDPI

Article

Estimating the Lifetime of Wireless Sensor Network Nodes through the Use of Embedded Analytical Battery Models

Leonardo M. Rodrigues [1],*, Carlos Montez [1], Gerson Budke [1], Francisco Vasques [2] and Paulo Portugal [2]

[1] Department of Automation and Systems, UFSC–Federal University of Santa Catarina, Florianópolis 88040-900, Brazil; carlos.montez@ufsc.br (C.M.); nandojve@gmail.com (G.B.)
[2] INEGI/INESC-TEC–Faculty of Engineering, University of Porto, Porto 4200-465, Portugal; vasques@fe.up.pt (F.V.); pportugal@fe.up.pt (P.P.)
* Correspondence: l.m.rodrigues@posgrad.ufsc.br; Tel.: +55-48-99188-4421

Received: 30 April 2017; Accepted: 13 June 2017; Published: 15 June 2017

Abstract: The operation of Wireless Sensor Networks (WSNs) is subject to multiple constraints, among which one of the most critical is available energy. Sensor nodes are typically powered by electrochemical batteries. The stored energy in battery devices is easily influenced by the operating temperature and the discharge current values. Therefore, it becomes difficult to estimate their voltage/charge behavior over time, which are relevant variables for the implementation of energy-aware policies. Nowadays, there are hardware and/or software approaches that can provide information about the battery operating conditions. However, this type of hardware-based approach increases the battery production cost, which may impair its use for sensor node implementations. The objective of this work is to propose a software-based approach to estimate both the state of charge and the voltage of batteries in WSN nodes based on the use of a temperature-dependent analytical battery model. The achieved results demonstrate the feasibility of using embedded analytical battery models to estimate the lifetime of batteries, without affecting the tasks performed by the WSN nodes.

Keywords: WSN; T-KiBaM; battery model; voltage modeling; thermal effect; micro-controller

1. Introduction

Wireless Sensor Networks (WSNs) are typically employed to support sensing/actuating activities in different application domains (e.g., industrial, commercial and residential) mainly due to their flexibility, low cost and low implementation complexity. A well-known constraint for the deployment of WSNs is the lifetime of their sensor nodes, which is upper-bounded due to stored energy limitations. The main limitation is the reduced battery capacity, which upper-bounds the operating life of the sensor node. In this context, it would be important to estimate both the battery State of Charge (SoC) and its lifetime according to the set of tasks executed by the nodes (e.g., data reception/transmission/processing tasks). For instance, this type of information is highly relevant, whenever energy-aware algorithms have to be implemented in the sensor nodes. However, estimating the battery lifetime in WSN nodes is a difficult task, as several factors influence their operation (e.g., chemical composition of the battery itself, operating temperature and discharge current) [1], resulting in a non-linear behavior over time [2–4].

There are two major options to estimate the battery operating behavior [5]: (i) hardware-based solutions, which involve the use of Integrated Circuits (ICs) that provide the relevant battery data; and (ii) software-based solutions, that usually require the use of adequate mathematical models. These two options are discussed below.

Smart batteries use ICs along with the electrochemical cell(s) to provide relevant data about the battery behavior (e.g., voltage, temperature, current) [6] and, in some cases, estimations about its operating behavior (e.g., SoC and remaining lifetime [7]) to the connected device (e.g., laptops, smartphones, cameras) [8,9]. However, the use of these hardware-based approaches increases the cost of producing batteries by approximately 25% (fuel gauge ICs costs about $2–3) [10]. In the context of WSNs, where the deployment of a large number of nodes may be required, such a solution may become economically infeasible. In addition, hardware-based solutions involving the use of ICs are often adapted to the integrated battery technology, where lookup tables are used to reconstruct the characteristics of the used cell(s) under different operating conditions [11]. Thus, it would be relevant to adopt software-based solutions able to accurately estimate the battery behavior of WSN nodes, without requiring the use of dedicated hardware. An important requirement is that, whatever the estimation approach, it must (i) be flexible enough to support different battery technologies; and (ii) present low computational cost due to the hardware constraints of sensor nodes.

Analytical battery models typically rely on a set of differential equations to estimate the battery behavior. Usually, these models are implemented in WSN simulators to estimate the operating behavior of the sensor nodes before their actual deployment. Within this context, the current battery condition is mathematically estimated to enable the deployment of energy-aware algorithms and protocols [12–14]. However, it is necessary to evaluate whether it is possible (or not) to implement similar differential equations-based models in real-world WSN nodes. It would be also necessary to assess the impact of implementing such mathematical models upon COTS low-power hardware. A pertinent question in this scenario is how does the computation of battery models may affect the lifetime of WSN nodes, which are usually based on low-power micro-controllers to save energy? In other words, and regarding the computational cost, is it feasible to perform a battery model computation upon a sensor node, in order to implement an on-line SoC determination and the related voltage level tracking functions?

The main target of this paper is to assess the usability of a low complexity analytical battery model [15], the Temperature-Dependent Kinetic Battery Model (T-KiBaM) [16], implemented upon Micro-Controller Units (MCUs) with low computational power, such as the ATmega328P and ATmega128RFA1 [17]. These MCUs are similar to those found in low-power COTS WSN nodes, e.g., the MICAz, which is based on the ATmega128L. Both ATmega-328P/-128RFA1 MCUs are widely available as the processing units of low cost WSN nodes. The main contributions of this work are:

- The experimental evaluation of a computationally inexpensive method to on-line estimate both the lifetime and State of Charge of batteries in real-world COTS WSN nodes with low computational capacity, small built-in memory and energy consumption constraints.
- A report on the implementation of light and accurate analytical battery models upon multiple low-power MCUs, typically used in real-world COTS WSN nodes.
- The implementation of a proof-of-concept application example demonstrating the usability of the T-KiBaM analytical model [16] to estimate the battery SoC and to on-line track its voltage level during the node activity period, as long as the discharge profile of the battery is known.

Next sections of this paper are organized as follows. Section 2 presents the related work. Section 3 introduces the basics about the T-KiBaM model, which includes the dependence of temperature on the estimation of the SoC of the battery, as well as a more accurate battery voltage model called Temperature-Dependent Voltage Model (TVM). Section 4 discusses the details about the experimental assessments and also about the model implementation. Section 5 presents the achieved results when running T-KiBaM upon low-power MCUs. Section 6 extends the previous section by adding a proof-of-concept application example, where other metrics are evaluated in an emulated operating scenario. Finally, Section 7 concludes the paper and presents future work.

2. Related Work

This section addresses the state-of-the-art in different areas of research related to this work. There are several studies dealing with the problem of estimating the SoC in batteries for different types of applications (e.g., electric vehicles) [18–21]. Nevertheless, the applicability of the reported results within the WSN context is difficult, as both the battery capacities and discharge profiles considered in these works are very different from those found in sensor nodes. Therefore, this section is divided in the two following items: (i) assessment of the computational cost of executing complex algorithms in micro-controllers with low computing capacity and (ii) deployment of analytical battery models in WSN nodes. Discussions on these topics are presented below.

There are several available studies in the literature evaluating the computational cost of running complex algorithms in low-power MCUs. For instance, Çakiroğlu [22] evaluated block cypher algorithms running upon an 8-bit Atmel ATmega128 MCU. The study assessed the execution of complex algorithms regarding the code/data memory requirements, execution time and throughput. Within the WSN context, Wei et al. [23] evaluated the overhead of cryptography algorithms suitable for WSNs. The authors evaluated the following metrics: clock cycles, code size, SRAM usage, and power consumption. The results showed that some algorithms are more appropriate when considering time-critical or energy-efficient applications, while others are more appropriate as they consume less SRAM memory. Capo-Chichi et al. [24] evaluated and compared the execution of data compression algorithms when using an ultra low-power micro-controller, known as MSP430, from Texas Instruments. The study aimed to evaluate the trade-off between energy consumption and compression efficiency. Guo et al. [25] presented two optimization approaches (Gauss-Newton Algorithm and Particle Swarm Optimization) to improve the localization of nodes in WSNs. The authors experimentally evaluated issues such as execution time, the number of iterations, memory usage and quality of estimation of the localization. Othman et al. [26] studied the cost of providing security in WSNs by implementing three cryptographic algorithms (AES, RC5 and RC6) upon MICA2 nodes. The authors analyzed memory consumption, operation time and energy consumption when using each one of the algorithms. Quirino et al. [27] presented the performance assessment of asymmetric cryptographic algorithms within the WSN context. Authors use three different platforms (ARM, MSP430 and AVR ATmega128) to evaluate the processing time of the algorithms. Pardo et al. [28] implemented an Artificial Neural Network (ANN) algorithm upon a low-cost chip (CC1110F32) for the purpose of developing autonomous intelligent WSNs to monitor and forecast the indoor temperature in smart homes. The authors were concerned with memory consumption and the use of computational resources, with the main objective of evaluating the feasibility of the implementation. Panić et al. [29] presented a micro-controller specifically designed to support WSN applications with severe security demands. The authors tested the developed chip with known cryptographic algorithms (ECC, AES and SHA-1), observing the execution time, security level and power consumption. Among all previously mentioned papers, note that the main evaluated metrics are: execution time, memory usage and power consumption. Some papers also evaluate other specific metrics related to the assessed algorithms, such as number of iterations and quality of the obtained results.

Regarding the implementation of analytical battery models upon WSN sensor nodes, Leveque et al. [30] presented a modeling approach to simulate the behavior of heterogeneous systems composed of WSN nodes. First, the authors model a set of WSN nodes for monitoring seismic perturbations, using a 32-bit microprocessor to solve a system of mathematical non-linear equations, that predict the battery behavior. This study case was implemented in SystemC-AMS, an extension of SystemC that models Analog/Mixed Systems. A computer simulates the system based on a 100-MHz microcontroller. In fact, this was a simulation-based assessment, with no hardware implementation. Other works also evaluated ways to estimate the battery lifetime on WSN nodes through simulations [6,31,32] or emulation [33]. Rahmé et al. [34] adapted the Rakhmatov and Vrudhula [35] analytical battery model to estimate the remaining energy in batteries of WSN nodes. The proposed battery model reduces the computational complexity and requires low memory usage. However, the achieved results illustrate relatively high errors (8–14%), when compared to the experimentally assessed results. Kerasiotis et al. [36] addressed the problem of estimating the battery lifetime on a WSN platform known as TelosB. In this case, the proposed methodology uses the energy consumption of each module to model the battery behavior of the node. The work used average load values to characterize the main operations performed at the node. The results indicated errors between 2–3% in comparison with experimental data for different duty cycles. Nataf and Festor [37] implemented the Rakhmatov and Vrudhula model in a dedicated operating system for WSN nodes known as Contiki. Tests included evaluations of node bootstrap and networking processes. In addition, the paper evaluated the accuracy of the lifetime estimate for different MAC protocols, also implemented in the Contiki operating system. Rukpakavong et al. [38] proposed a dynamic approach that considers several factors which can influence the battery lifetime, such as self-discharge, aging, discharge current and temperature. This approach was implemented in two WSN platforms: MICA2 and N740 NanoSensor. The results indicated deviations from −3.5% to 2.5% when estimating the battery lifetime. However, it was assumed that the voltage value must be read at the beginning of the calculations to evaluate the initial battery capacity, i.e., this approach did not track the voltage of the battery over time. In addition, the work did not consider the recovery effect, which is an important effect in scenarios where the WSN nodes operate in duty cycle scheme.

The main advantage of the methodology proposed in this paper is related to the deployment of a temperature-dependent battery model, which can be used to predict the behavior of the battery in low-power WSN nodes regardless of the associated hardware. The proposed methodology assumes that there is a cyclical operation pattern for the WSN nodes (e.g., a duty cycle), so that the discharge profile can be used as input parameter to compute the battery behavior over time (open-loop computation). Figure 1 depicts an example of a discharge profile based on a MICA2DOT WSN node [39]. By using this type of discharge profiles, it becomes possible to obtain two information about the battery: (i) the SoC, which is obtained through the analytical battery model proposed in Section 3.2; and (ii) the voltage level, which is concurrently obtained through the execution of the voltage model presented in Section 3.2.1.

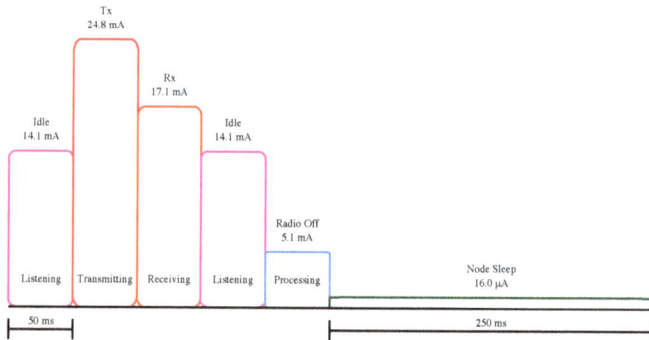

Figure 1. Example of a discharge profile based on [39]. Tx = Transmitting; Rx = Receiving.

3. Background

This section presents the main concepts involved in this paper. Briefly, the original Kinetic Battery Model (KiBaM) [40,41] model is presented along with its voltage model. Next, it is introduced a summary on the Temperature-Dependent Kinetic Battery Model (T-KiBaM) [16], an extension of the KiBaM model that includes the effect of temperature on the predictions about both the lifetime and voltage of the battery.

3.1. Kinetic Battery Model (KiBaM)

KiBaM is one of the first high-accuracy analytical battery models that was proposed in the early nineties. It is based on an intuitive approach to model the behavior of high-capacity Lead-Acid batteries over time. This model uses a two-tank analogy to describe the battery charge and discharge processes, as shown in Figure 2.

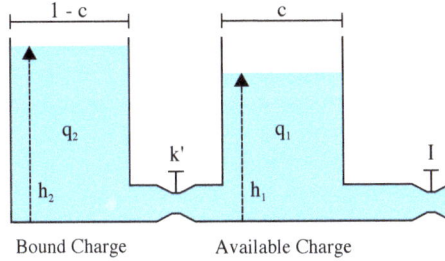

Figure 2. Kinetic Battery Model (KiBaM) (based on [40]).

In this model, the available charge tank is the power supply for any device that consumes a current over time, $I(t)$. The average value of current I is considered for each period of time t. The bound charge tank holds part of the battery charge, which can be transposed to the available charge tank at a rate k' through a valve that interconnects both thanks. In this context, k' is a constant that represents the rate of a chemical diffusion/reaction process. The transfer of charge occurs as long as there is a height difference between the charges of both tanks, i.e., $\delta = h_2 - h_1 \neq 0$. The constant c indicates the total charge ratio stored in the available charge tank. The battery remains operational as long as there is charge in the available charge tank (i.e., SoC > 0%), regardless of whether there is charge in the bound charge tank or not. A system of differential equations is able to describe the KiBaM model (refer to [40] for further details). Laplace transforms are able to solve such a system of differential equations, resulting in:

$$
\begin{cases}
q_1 = q_{1,0} \cdot e^{-k \cdot t} + \dfrac{(q_0 \cdot k \cdot c - I) \cdot (1 - e^{-k \cdot t})}{k} - \dfrac{I \cdot c \cdot (k \cdot t - 1 + e^{-k \cdot t})}{k} \\[4mm]
q_2 = q_{2,0} \cdot e^{-k \cdot t} + q_0 \cdot (1 - c) \cdot (1 - e^{-k \cdot t}) - \dfrac{I \cdot (1 - c) \cdot (k \cdot t - 1 + e^{-k \cdot t})}{k},
\end{cases}
\tag{1}
$$

where $q_{1,0}$ and $q_{2,0}$ are the amount of charge in the available and bound charge tanks, respectively, when $I = 0$. A new rate constant is defined as $k = k'/(c \cdot (1 - c))$. In addition, $q_0 = q_{1,0} + q_{2,0}$, where q_0 is the amount of charge in the battery at $t = 0$. Thus, the constants required for the use of KiBaM are: q_{max} (the maximum capacity of the battery), c (a fraction of the capacity stored in the available charge tank) and k (the rate constant). Such constants can be obtained from discharge tests with real batteries, as presented in [40], or applying the data-sheet values when available, and used in analytical evaluations to determine the SoC of the battery.

The SoC of the battery can be calculated from the relation of the unavailable charge of the battery, $q_{un}(t) = (1 - c) \cdot \delta(t)$, and also from Equation (1) along with the discharge current, $i(t)$, as follows [4]:

$$
SOC(t) = SOC_{initial} - \frac{1}{q_{max}} \left[\int i(t)\, dt + q_{un}(t) \right].
\tag{2}
$$

3.1.1. KiBaM Voltage Model

The earlier KiBaM model is also able to track the battery voltage (V) over time. To do so, it becomes necessary to expand the battery model with the related electrical model:

$$V = E - I \cdot R_0, \tag{3}$$

where R_0 is the internal resistance of the battery and E is the internal voltage of the battery. For discharge purposes, E can be obtained using the following equation:

$$E = E_{min} + (E_{0,d} - E_{min})\frac{q_1}{q_{1,max}}, \tag{4}$$

where E_{min} is the minimum allowed internal discharge voltage ("empty"), $E_{0,d}$ is the maximum internal discharge voltage ("full") and $q_{1,max}$ is the maximum capacity of the available charge tank (obtained from q_{max}). The internal resistance, R_0, can be experimentally determined using constant discharge currents. Its value is represented by the slope dV/dI, when the battery is fully charged. In other words, the slope of $V \times I$ gives the value for R_0 [40].

Note that this voltage model is quite limited for the most commonly used battery technologies within the WSN context (e.g., Ni-MH or Li-ion), as it assumes a linear behavior when the battery is discharged with a constant current. A more accurate solution will be presented in the next section.

3.2. Temperature-Dependent Kinetic Battery Model (T-KiBaM)

T-KiBaM [16] is an extension of the KiBaM model that aggregates the effects caused by the use of the battery at different temperatures, which may change both its lifetime and its voltage behavior over time. Briefly, the thermal effect can accelerate the rate of reactions inside the battery, implying that the battery can provide a higher effective capacity at high temperatures [42]. The influence of temperature on the rate of a chemical reaction follows an empirical law known as the Arrhenius equation:

$$k = A \cdot e^{-\frac{E_a}{R \cdot T_k}}, \tag{5}$$

where k is the constant rate of a reaction, A is the pre-exponential factor or pre-factor (in s^{-1}), E_a is the activation energy (in KJ/mol), R is the universal gas constant (8.314×10^{-3} KJ/mol·K) and T_k is the temperature (in Kelvin). Considering that both k parameters from KiBaM and the Arrhenius equation refer to a constant reaction rate, it becomes possible to establish the following relationship: $k_{KiBaM} = k_{Arrhenius}$. Therefore:

$$k_{KiBaM} = A \cdot e^{-\frac{E_a}{R \cdot T_k}}. \tag{6}$$

That is, the parameter k of the KiBaM model now follows the Arrhenius relation, and may vary according to the operating temperature of the battery. As described by Rodrigues et al. [16], a reduced number of experimental measurements is required to determine the values of constants A and E_a.

In addition, temperature also influences the charge capacity provided by the battery. Typically, batteries provide higher effective capacities at higher temperatures [42] and lower effective capacities when used at lower temperatures [43]. Within this context, it is crucial to adjust T-KiBaM to the technology of the battery being modeled (e.g., Ni-MH or Li-ion). Briefly, it is necessary to experimentally observe the behavior of the battery at different temperatures and discharge currents. Then, it is possible to establish a Correction Factor (CF), which allows the creation of a function capable of correcting the initial capacity of the battery according to the temperature in Celsius degrees (T_c). Please refer to [16] for details on how to find a function that allows calculating the CF value.

3.2.1. T-KiBaM Voltage Model

T-KiBaM also includes its own Temperature-Dependent Voltage Model (TVM) [16], which is an extension of the Tremblay–Dessaint voltage model [44,45]. TVM is able to provide specific voltage curves ($V \times t$) for different operating temperatures of the battery. The advantage of this approach is related to the accuracy of the voltage curve when compared to experimental results. Equation (7) describes the behavior of the voltage curve for the discharge of Ni-MH batteries (although the battery voltage is a function of time, the representation $V_b(t)$ is not used for simplification purposes).

$$V_b = E_0 - R_b \cdot i - K_b \cdot \frac{Q}{Q - it \cdot \tau_b} \cdot (it \cdot \tau_b + i^*) + Exp(t), \tag{7}$$

where V_b is the battery voltage (V), E_0 is the battery constant reference voltage (V), R_b is the internal resistance (Ω), K_b is the polarization resistance (Ω), Q is the battery capacity (Ah), $it = \int idt$ is the actual battery charge (Ah), i is the battery current (A) and i^* is the filtered current (A). For further details, please refer to [44,45]. It is also well-known that Nickel-based batteries exhibit a hysteresis phenomenon between the charge and discharge processes, which occurs only at the beginning of the discharge curve, regardless of their SoC. This phenomenon can be represented by a non-linear dynamic system:

$$\dot{Exp}(t) = \tau_b \cdot B \cdot |i(t)| \cdot (-Exp(t) + A_b \cdot u(t)), \tag{8}$$

where B is the exponential zone time constant inverse (Ah)$^{-1}$, $i(t)$ is the battery current (A), $Exp(t)$ is the exponential zone voltage (V), A_b is the exponential zone amplitude (V) and $u(t)$ is the charge/discharge mode. The exponential voltage relies on its initial value $Exp(t_0)$ and the charge ($u(t) = 1$) or discharge ($u(t) = 0$) mode. A smoothing constant (τ_b) has been added in order to increase the accuracy of the Tremblay–Dessaint voltage model. Thus, the parameters required to model different battery types are as follows: E_0, R_b, K_b, A_b, B and τ_b. Such parameters can be obtained from the battery datasheet or through a set of simple experimental measurements [45].

In this case, the parameters of the TVM model are obtained through experiments at different temperatures, which allows to use the Arrhenius equation to relate the influence of temperature on the behavior of the parameters. In other words, it becomes possible to obtain the values of the parameters at different temperatures (cf. Table 5 of [16]). Thus, the TVM model is able to provide the battery voltage level at any instant of time, regardless of the considered ambient temperature. Table 1 illustrates the set of parameters used to model a Ni-MH battery (Panasonic HHR-4MRT/2BB).

Table 1. T-KiBaM parameters for a Ni-MH battery [16].

Model	Parameter	Value
	E_a	1.1949
Arrhenius	A	0.96397
	R	0.008314
CF (T_c)	a, b, c, d	cf. Table 4 of [16]
	c	0.56418
T-KiBaM	k	$A \cdot e^{\frac{-E_a}{R \cdot T_k}}$
	y_0	$750 \cdot$ CF (T_c)
TVM	$A_b, B, E_0, Exp(t_0),$ K_b, Q, R_b, τ_b	cf. Table 5 of [16]

4. T-KiBaM Implementation

The main target for the use of the temperature-dependent battery models presented in the previous section is for the prediction of the lifetime and voltage behavior of typical WSN batteries. Basically, these are analytical models that can be included in simulation models, to perform the simulation assessment of WSN deployments. On the other hand, one of the main targets of this paper is to show that these models can be also implemented upon low-power small-memory MCUs, providing accurate results for the prediction of both the lifetime and the voltage behavior of typical

WSN batteries. Therefore, the purpose of this section is to present a set of T-KiBaM functions, which may be deployed upon WSN-compatible MCUs. These software functions will be experimentally validated by comparing their analytical and experimental results.

4.1. T-KiBaM Functions

The purpose of T-KiBaM is to provide an estimate of the State of Charge (SoC) of the battery over time at different temperatures, including information about its voltage level. Therefore, it becomes possible to obtain the estimated battery lifetime according to the discharge profile and the used temperature. The implementation presented in this work is divided into two stages: (i) the call to the T-KiBaM function and (ii) the T-KiBaM function itself. Such stages are described below.

The first stage implements the call to the T-KiBaM function, that has as input the discharge profile. Such discharge profile (DP) is defined by a set of pairs (I_x, t_{I_x}), where I_x represents the discharge current and t_{I_x} represents its operating time (or time step), with $x = 1, 2, 3, \ldots, n$. For example, $DP_{set} = [(I_1, t_{I_1}); (I_2, t_{I_2}); \ldots; (I_n, t_{I_n})]$. Therefore, this stage returns the updated values regarding the T-KiBaM and TVM functions. Algorithm 1 shows the implementation of the function call that uses Equations (1) and (7) to update the battery data.

Algorithm 1: T-KiBaM_call.

    Input: $E_a, A, R, T, q_0, c, k, t_0, DP_{set}$,
            $E_0, R_b, K_b, \tau_b, B, prExp$
    Output: q_1, q_2, t_0, V_b
1   $q_0 = q_0 \cdot CF(T)$;
2   $q_1 = (c) \cdot q_0$;
3   $q_2 = (1 - c) \cdot q_0$;
4   $k = A \cdot e^{-Ea/(R \cdot T)}$;
5   $It = 0$;
6   **foreach** $(I_x, t_{I_x}) \in DP_{set}$ **do**
7       $It = It + (I_x \cdot t_{I_x})$;
8   **if** $q_1 > 0$ **then**
9           $[q_1, q_2, t_0]$ = T-KiBaM_function $(c, k, q_1, q_2, t_0, I_x, t_{I_x})$;
10          $Exp = (1/(1 + (B \cdot I_x \cdot t_{I_x} \cdot \tau_b))) \cdot prExp$;
11   V_b = TVM_function $(E_0, R_b, K_b, \tau_b, B, q_0, I_x, It, Exp)$;
12          $prExp = Exp$;
13   **end**
14 **end**
15 **return** (q_1, q_2, t_0, V_b);

The input parameters at this stage are related to the Arrhenius equation (E_a, A, R, T), to T-KiBaM (q_0, c, k, t_0, DP_{set}) and TVM (E_0, R_b, K_b, τ_b, B, $prExp$). In T-KiBaM parameters, note that q_0 represents the initial battery capacity. In this case, this parameter receives the nominal capacity of the battery used as reference. The parameter values of c and k are dependent on the battery technology. In this work, these three values were obtained from a Panasonic battery, model HHR-4MRT/2BB (2xAAA, 2.4 V, 750 mAh). For more information on how to obtain these parameters, please refer to [16]. Next, parameter t_0 represents the total battery lifetime. In addition, parameter DP_{set} may contain one or more pairs (I_x, t_{I_x}) to indicate the use of a set of tasks (i.e., a discharge profile as depicted in Figure 1). This feature is useful as a WSN node usually has different discharge currents for different operating states, e.g., Tx, Rx and Sleep. Using the DP_{set} definition, duty cycles can also be used in the T-KiBaM implementation. In the TVM parameters, $prExp$ represents the initial value of the exponential voltage, $Exp(t_0)$, which is used for the calculation of $Exp(t)$ in each iteration.

In Algorithm 1, the correction factor (CF) function is applied in Line 1, as described in Section 3.2. In addition, the definition of k (Line 4) considers the Arrhenius equation values (E_a, A, R and T), which can be obtained through experiments (please refer to [16] for details). Through the for loop (Line 6), it becomes possible to call the T-KiBaM function according to the used discharge profile. As presented in Line 8, the user of T-KiBaM should check the content of the available charge tank, which needs to be greater than zero. This is a necessary condition for the battery operation, even if there is charge at the bound charge tank. Finally, note that the battery voltage level is obtained in Line 11, which performs the calculations corresponding to Equation (7). Finally, the algorithm returns some additional information about the battery, such as remaining battery charge in both tanks (q_1 and q_2), battery run time (t_0), and voltage level (V_b) when executing the discharge profile DP_{set}.

The T-KiBaM function implements the concepts presented in Section 3.1, where Equation (1) is used to calculate the charge of the battery over time. This stage returns the updated values in relation to the battery charge and its time of use. Algorithm 2 shows how to implement the T-KiBaM function.

Algorithm 2: T-KiBaM_function.

Input: $c, k, q_{1,0}, q_{2,0}, t_0, I, t_I$
Output: q_1, q_2, t
1 $q_0 = q_{1,0} + q_{2,0}$;
2 $t = t_0 + t_I$;
3 $q_1 = \text{compute-i } (c, k, q_0, q_{1,0}, q_{2,0}, I, t_I)$;
4 $q_2 = \text{compute-j } (c, k, q_0, q_{1,0}, q_{2,0}, I, t_I)$;
5 **return** (q_1, q_2, t);

The input parameters of the T-KiBaM function are as follows: $c, k, q_{1,0}, q_{2,0}, t_0, I$ and t_I. The values of I and t_I represent a task in the DP_{set}. Note that Lines 3 and 4 perform the calculations corresponding to Equation (1). The output values of q_1 and q_2 represent the actual state of charge in the available and bound charge tanks, respectively. Finally, t represents a time accumulator that is used to compute the total time of battery usage.

The knowledge about the SoC of the battery is very important for the development of energy-aware strategies. In this approach, during the node duty cycle, for example, it is possible to perform an iteration of T-KiBaM for each performed task (e.g., Tx, Rx, Sleep) in order to update the battery status (SoC and voltage level). With this, the node can take different decisions according to the current capacity of the battery. Although the proposed approach is flexible in several aspects, the following assumptions should be considered when running the T-KiBaM model:

1. The node initializes its operating cycle with a fully charged battery, i.e., SoC = 100%. In addition, the T-KiBaM model is adjusted for the used battery technology. Therefore, it is not necessary to measure any battery information over time (e.g., voltage level);
2. The node knows the discharge profile for all tasks that need to be performed during its operation. Knowing the discharge current in the transition between states, as well as the time it takes to perform such action, makes the T-KiBaM even more accurate. Thus, it is possible to parametrize T-KiBaM with the measured values and the time spent in each state/transition. In this case, the better the discharge profile definition, the greater the accuracy of the estimated battery behavior. Note that the discharge profile can be obtained from an analysis of the hardware power consumption (e.g., MCU, transceiver, sensors, etc.);
3. The duty cycle of the node does not have to be constant since T-KiBaM supports different operating times (t_{I_x}) for each task (I_x), allowing the configuration of any combination of tasks;
4. The node can obtain the environment temperature, which is provided to the T-KiBaM model to increase the accuracy of the estimate on the battery behavior.

An application example is presented in Section 6 to demonstrate the use of the T-KiBaM model.

4.2. Analytical vs. Experimental Comparison

The objective of this section is to validate the analytical evaluations comparing analytical results obtained from the T-KiBaM model with some experimental results, comparing the error between the two approaches regarding the battery lifetime estimation and its voltage behavior over time. Note that the values of all the constants of the T-KiBaM model were previously obtained by Rodrigues et al. [16]. In addition, all the analytical evaluations use the same experimental characteristics, such as discharge profile and temperature.

First, tests with continuous discharge currents were performed at different temperatures to evaluate the accuracy of the T-KiBaM model. The evaluated temperatures were as follows: -5, 10, 25, 32.5, and 40 °C. The used discharge currents were 20 and 30 mA. With this, it became possible to analyze the relative Error (ERR) between analytical and experimental results. Table 2 presents the results of these evaluations. The experimental results (EXP), T-KiBaM and ERR columns represent, respectively, the experimental average lifetime of three battery measurements (note that the cutoff value of 2.0 V is considered for the calculation of the battery lifetime), the lifetime using T-KiBaM (in this case, the lifetime is reached when SoC = 0%) and the relative error between EXP and T-KiBaM. The average ERR (AVG) values are presented at the of the table.

Table 2. Battery lifetime *. EXP: Experimental result; T-KiBaM: analytical result; ERR: relative Error.

	-5 °C			10 °C			25 °C			32.5 °C			40 °C		
I (mA)	EXP (h)	T-KiBaM (h)	ERR (%)	EXP (h)	T-KiBaM (h)	ERR (%)	EXP (h)	T-KiBaM (h)	ERR (%)	EXP (h)	T-KiBaM (h)	ERR (%)	EXP (h)	T-KiBaM (h)	ERR (%)
20	36.714	36.866	0.41	37.402	37.361	0.11	37.984	37.814	0.45	37.835	37.976	0.37	37.133	37.271	0.37
30	24.749	24.750	0.00	25.087	25.082	0.02	25.385	25.386	0.00	25.560	25.495	0.26	25.022	25.022	0.00
AVG			0.20			0.06			0.22			0.31			0.18

* Results using continuous discharge currents.

Next, some experiments using a Duty Cycle (DC) scheme were also carried out to evaluate the ability of the T-KiBaM model to handle typical WSN scenarios. The discharge current was set at 30 mA to decrease the time of the experiments. The following duty cycle schemes were evaluated:

$$DC_{75\%} = [(I_1 = 30 \text{ mA}, t_{I_1} = 3 \text{ s}); (I_2 = 0.0 \text{ mA}, t_{I_2} = 1 \text{ s})];$$
$$DC_{50\%} = [(I_1 = 30 \text{ mA}, t_{I_1} = 1 \text{ s}); (I_2 = 0.0 \text{ mA}, t_{I_2} = 1 \text{ s})];$$
$$DC_{25\%} = [(I_1 = 30 \text{ mA}, t_{I_1} = 1 \text{ s}); (I_2 = 0.0 \text{ mA}, t_{I_2} = 3 \text{ s})].$$

Note that the duty cycle period is 4 s for $DC_{75\%}$ and $DC_{25\%}$, and 2 s for $DC_{50\%}$. In addition, only the temperature at 25 °C was used in the experiments. Table 3 presents the results of this evaluation, including the relative Error for each situation.

Table 3. Battery lifetime *. EXP: Experimental result; T-KiBaM: analytical result; ERR: relative Error.

	25 °C		
Duty Cycle (%)	EXP (h)	T-KiBaM (h)	ERR (%)
75	33.524	33.849	1.81
50	51.229	50.774	0.89
25	102.547	101.549	2.49
AVG			1.73

* Results using duty cycle schemes.

These results demonstrate that T-KiBaM is able to accurately estimate the battery lifetime of WSN nodes, presenting average ERR values smaller than 0.35% for continuous discharge currents and an average ERR value of 1.73% for duty cycle schemes. However, although the presented results are quite accurate, battery lifetime is not the only interesting information that can be extracted from the T-KiBaM model.

The voltage level is another relevant factor when evaluating the behavior of batteries. In the case of T-KiBaM, the battery voltage model provides voltage values that are dependent on the operating temperature, which allows monitoring the state of the battery more accurately, particularly, in WSN scenarios with high temperature variations. Figure 3 depicts an example comparing the experimental results using a continuous discharge current (30 mA) at different temperatures with those analytically obtained using the T-KiBaM and KiBaM models. The experimental data are fitted according to the average behavior of three experiments.

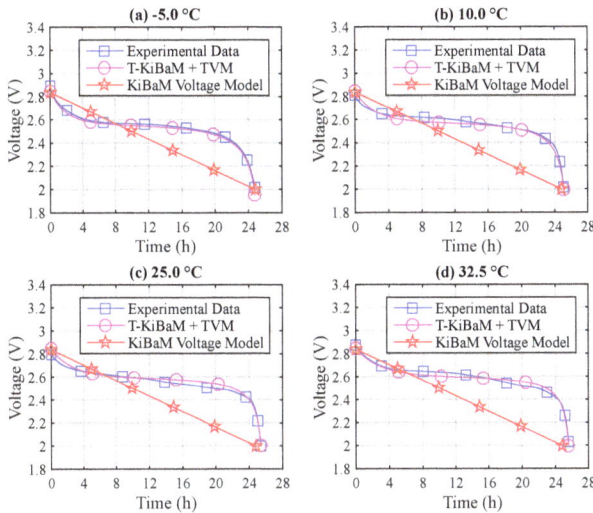

Figure 3. Voltage tracking comparison using a constant discharge current (30 mA).

Note that the original KiBaM voltage model represents a linear battery discharge curve, $V \times t$. This type of approximation induces significant errors with respect to the lifetime analysis of any device connected to the battery. On the other hand, the T-KiBaM + TVM model offers a higher precision when estimating the behavior of the battery voltage curve over time. For instance, at $T = -5\ °C$ (Figure 3a), analyzing the voltage level at 2.4 V, the relative error to the experiment of KiBaM is 37.53%, while in T-KiBaM is 0.73%.

5. Running T-KiBaM in Low-Power MCUs

This section presents the experimental results obtained when implementing the T-KiBaM model in multiple WSN-compatible MCUs. The objective is to check if analytical battery models, embedded in a low computational capacity hardware, can be used to track both the battery SoC and the voltage level of the battery itself over time. First, we present the basic characteristics of each MCU used in this work. Then, a discussion is included regarding the selected metrics used for the experimental evaluations. Finally, the results obtained from the experimental assessment are presented.

5.1. MCUs and Related Hardware Platforms

Arduino (https://www.arduino.cc) is an open-source platform that has been designed to facilitate electronic circuits prototyping. Arduino boards support the addition of sensors and/or actuators to existing designs, allowing the interaction with the physical environment. The use of this platform is highly popular due to its low cost, compatibility between operating systems, as well as the easy extensibility of both software and hardware. There are multiple Arduino board types. This work focuses on the UNO version that includes an Atmel ATmega328P low-power AVR 8-bit microcontroller, which has 32 KB of integrated Flash memory, as well as 2 KB of SRAM and 1 KB of EEPROM. This MCU operates at 16 MHz on the UNO board. The current consumption at 1 MHz is 0.2 mA in active mode [46]. Other MCUs were also used in the experimental assessments. These MCUs are using C code with specific manufacturer library (http://www.atmel.com/tools/avrsoftwareframework.aspx). The specifications of each used micro-controller are summarized in Table 4.

Table 4. Specifications of the used MCUs.

MCU	Platform	Clock (MHz)	Wait State	FPU	Flash (KB)	SRAM (KB)	EEPROM (KB)	Typical Current (mA)
ATmega328P	8-bit AVR	16	0	no	32	2	1	0.2
ATmega128RFA1	8-bit AVR	16	0	no	128	16	4	4.1
ATxmega256A3U	8/16-bit AVR	32	0	no	256	16	4	9.5
SAMR21G18A	32-bit ARM Cortex-M0+	48	1	no	256	32	0	6.7
SAMG55	32-bit ARM Cortex-M4	120	5	yes	512	160	0	24.2
SAMV71Q21	32-bit ARM Cortex-M7	300	6	yes	2048	384	0	83.0

The Atmel ATmega128RFA1 is an 8-bit AVR MCU, which has a built-in 128 KB of Flash memory, as well as 16 KB of SRAM and 4 KB of EEPROM. The MCU can operate up to 16 MHz [47]. The Atmel ATxmega256A3U is an 8/16-bit AVR XMEGA low-power MCU that features 256 KB of Flash memory, as well as 16 KB of SRAM and 4096 bytes of EEPROM. This MCU can run at 32 MHz [48]. The Atmel SAMR21G18A MCU uses a low-power 32-bit ARM Cortex-M0+ processor. This chip has a 256 KB of Flash memory, plus 32 KB of SRAM [49]. The Atmel SAMG55 is based on the ARM architecture. This MCU has a 32-bit Cortex-M4 core that can reach speeds up to 120 MHz with a Floating Point Unit (FPU). In addition, this chip has 512 KB Flash Memory and 160 KB SRAM plus up to 16 KB (cache + I/D RAM) [50]. The Atmel SMART SAMV71Q21 is based on the ARM architecture, featuring a Cortex-M7 RISC 32-bit processor with a FPU. This MCU can reach speeds up to 300 MHz, featuring 2048 KB of Flash memory, as well as a dual 16-KB cache and 384 KB of SRAM memories [51].

5.2. Performance Metrics

As the results presented in Section 4 are consistent with those found in the experimental assessment, i.e., T-KiBaM parameters have been properly adjusted, a computer with a 2.9 GHz Intel Core i5 processor running MATLAB is used as the basis of the comparisons regarding the battery lifetime estimation. This is considered the best platform for the execution of this algorithm as it presents an interesting precision regarding the number of significant figures. Thus, the tested set includes the same experimental continuous discharge currents, as well as a variety of other discharge current values. Such set comprises the following currents: 5, 10, 20, 30, 40, 50, 60, 70, 80, 90, and 100 mA.

The following metrics are used for this experimental assessment: (i) algorithm execution time; (ii) memory usage; (iii) energy consumption; (iv) number of iterations of the algorithm for different tasks; and (v) estimated battery lifetime.

5.3. Experimental Results Using Low-Power MCUs

The results shown in this section were obtained by running the T-KiBaM functions on different low-power MCUs. Note that, when using continuous discharge currents in the analytical evaluations, the T-KiBaM function needs an operating time t_1 (or time step) as input to run the battery model. Hereafter, a 1-second step was assumed between consecutive executions as it represents a relevant low granularity when continuous discharge currents are used to feed the model, if compared to the total battery discharge time. A discussion regarding the time step size is performed in Section 6.3.

5.3.1. Execution Time

The first evaluated metric is the function execution time (ET) when running T-KiBaM in low-power MCUs. The objective is to compare the performance of the algorithm in platforms with different characteristics in order to verify the possibility of its implementation in WSN nodes.

It is important to note that the results presented in this section consider the average of three executions of the algorithm. The execution times were collected from checkpoints at the beginning and at the end of the T-KiBaM function call. In addition, all micro-controllers can only access the flash memory with a maximum clock of 32 MHz and, after that speed, wait-states must be inserted. All the performed experiments used the best configuration to achieve the fastest results. Note that the focus of this work is not on the evaluation of the faster micro-controller, therefore, the source code was compiled with -02 option and no specific optimization was performed in the available libraries. The FPU has been enabled in all MCUs with this option. The instruction cache has been enabled in SAMG55 and the instruction/data cache have been enabled in SAMV71. The use of same MCU manufacturer allowed both to unify code and test the same library for all MCU models. Table 5 presents the average execution times achieved by each platform.

Table 5. Execution times (average) on all platforms *.

	ATmega 328P	ATmega 128RFA1	ATxmega 256A3U	SAMR 21G18A	ATSAM G55	SAMV 71Q21
Execution Time (µs)	549.02	499.86	259.10	1311.65	164.87	5.33

* Results using the clock frequencies shown in Table 4.

The results point to average execution times of less than 1.4 ms on all platforms. The SAMV71Q21 micro-controller presented an average execution time close to 5.3 µs. This result is within the expected range, since this MCU operates at a higher frequency, i.e., 300 MHz. On the other hand, the SAMR21G18A micro-controller delivers a poor performance for a MCU from its category. The average execution time around 1.3 ms, even when operating at 48 MHz, could be related to lack of code optimization of GCC compiler [52] that increases code size and, consequently, slows down the code execution considerably. The performed experiments clarified that optimization should be mandatory to achieve better results. Tests also have shown that ARM and AVR produce similar results when using soft float ABI (Application Binary Interface) and no cache since ARMs, probably, are stalled waiting for new instruction due to wait-states. Despite this, the rational indicates that the obtained values are feasible when compared with real-world applications, such as the use of encryption algorithms in WSNs with low-power MCUs [26].

5.3.2. Memory Usage

The second evaluated metric is memory usage. Analyzing the amount of Flash memory occupied by the T-KiBaM model is an important metric, as micro-controllers used in WSN nodes usually have very little available memory. In this sense, it is possible to establish the spatial cost of implementing an analytical battery model in a low-power MCU.

Note that the results presented in this section consider only the memory usage relative to the T-KiBaM model source code implementation and the essential compile components on each platform. In other words, libraries and debugging codes are not considered in this analysis. Table 6 presents the memory usage on all platforms, including the percentage of total available memory.

Table 6. Memory usage on all platforms.

	ATmega 328P	ATmega 128RFA1	ATxmega 256A3U	SAMR 21G18A	ATSAM G55	SAMV 71Q21
Memory usage (KB)	7.444	11.384	19.254	40.376	39.136	25.712
Total Available (KB)	32	128	256	256	512	2048
Percentage of total (%)	23.0	8.9	7.5	15.7	7.6	1.2

According to Table 6, the implementation of T-KiBaM on the SAMR21G18A occupies approximately 40.3 KB, the highest memory occupancy among all platforms. On the other hand, the ATmega328P presents the lowest memory occupancy, with only 7.4 KB. However, in relation to the total Flash memory availability, this micro-controller has the highest occupancy, about 23.0% of 32 KB in total. The SAMV71Q21 has the lowest memory occupancy rate in percentage terms.

As observed in Table 6, three of the five tested platforms have memory occupancy rates of less than 10%. Thus, these results show that it is feasible to implement an analytical battery model on a low-power WSN node, such as the iLive node [53] which features 128 KB of Flash memory.

5.3.3. Power Consumption

The power consumption is the third metric evaluated in this work. The objective is to evaluate how much energy consumes an iteration of the T-KiBaM algorithm. For this, it is necessary to measure the current consumed by each MCU first. Further details are given below.

A multimeter (MD-6450 True-RMS) was used to measure the current on each platform. All measurements were taken with the board of each micro-controller connected via USB while running the T-KiBaM model. Voltage variations are not considered since the algorithm execution time is very small (<1.4 ms). Thus, the average values for voltage ($\approx$5.05 V) and current are considered in the calculations of this section. Table 7 shows the measured current values as well as the electrical power for each micro-controller, calculated through the relation $P = V \times I$.

Table 7. Power consumption in each platform.

	ATmega 328P	ATmega 128RFA1	ATxmega 256A3U	SAMR 21G18A	ATSAM G55	SAMV 71Q21
Power Supply (V)	5.05	5.05	5.05	5.05	5.05	5.05
Current (mA)	49.3	77.1	18.9	12.0	30.2	82.6
Power (mW)	248.9	389.3	95.4	60.6	52.5	417.1

From these results, it is possible to obtain the energy spent according to the execution time of an iteration of T-KiBaM algorithm in each micro-controller, through the relation $E = P \times \Delta t$. In this case, Δt is obtained from the execution time in each platform. Therefore, the energy spent is directly related to the first metric, the execution time. Table 8 shows the average energy spent when running a single iteration of T-KiBaM on each platform.

Table 8. Energy spent (average) on a single iteration of the algorithm on all platforms.

	ATmega 328P	ATmega 128RFA1	ATxmega 256A3U	SAMR 21G18A	ATSAM G55	SAMV 71Q21
Energy Spent (mJ)	0.1366	0.1945	0.0247	0.0794	0.0086	0.0022

6. T-KiBaM Usage in WSN Nodes: Application Example

WSN nodes usually perform several tasks during their operation, including data transmission (Tx), reception (Rx) and processing (Pr). It is also possible to save energy during certain intervals of time by putting the nodes in sleep mode (Sl). Generally speaking, such nodes operate in duty cycle scheme, i.e., cyclically repeating a sequence of tasks over time, until their battery power runs out. The objective of this section is to illustrate the usage of T-KiBaM in a real application, considering the operating characteristics of real WSN nodes. With this, other performance metrics can be assessed in relation to the execution of the T-KiBaM model in low-power MCUs. Finally, the presented application example is used in a sensitivity analysis, where variations are applied to the input parameters of the T-KiBaM model.

6.1. Application Example

The application scenario described in this section cover the mode of operation of most WSN applications. In this sense, two scenarios are described: (i) the node remains 100% of the time in the active mode; and (ii) the node operates in a duty cycle scheme, i.e., inserting periods in sleeping mode alternately with its active period. Further details are given below.

A set of tasks (discharge profile) can be used to properly emulate the operation of the nodes, i.e., discharge the battery charge when performing different tasks. However, to simplify the analysis, it is assumed that the node performs only one useful task (e.g., Rx, Tx, or Pr) in both scenarios. A task is defined by the discharge current and its operating time (I, t_1), including periods in sleeping mode. The node executes the T-KiBaM algorithm at the end of its task to update the state of charge of its battery. Although it may play a significant role in energy consumption, the node initialization process is not considered in these analysis, since it runs only once during its entire life cycle. Figure 4 depicts a schematic summarizing the node activities in the two presented scenarios.

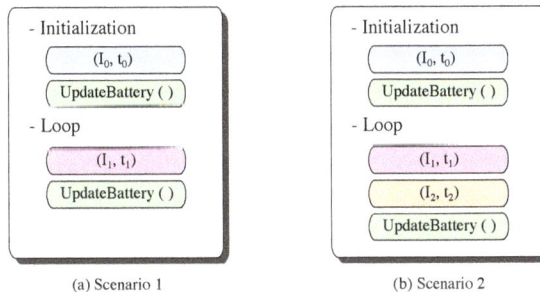

- Initialization
  (I_0, t_0)
  UpdateBattery ( )
- Loop
  (I_1, t_1)
  UpdateBattery ( )

(a) Scenario 1

- Initialization
  (I_0, t_0)
  UpdateBattery ( )
- Loop
  (I_1, t_1)
  (I_2, t_2)
  UpdateBattery ( )

(b) Scenario 2

Figure 4. Node activity description. (**a**) Active mode only; (**b**) Active and inactive modes.

As depicted in Figure 4a, Scenario 1 presents the operating mode of a node operating 100% of the time in active mode. Note that the main loop considers only the performed task, represented by (I_1, t_1), and the update of the battery state of charge and voltage level. On the other hand, Figure 4b presents Scenario 2, which adds a sleep mode period, represented by (I_2, t_2), at each duty cycle. In this sense, the node performs its main task, goes into low-power mode (Sl), and then updates the battery state of charge and voltage level.

6.2. Estimating the Battery Lifetime

The fourth metric assessed in this work is the battery lifetime estimation. One of the main features of T-KiBaM model is to provide the estimated battery lifetime according to the used discharge profile. Therefore, a modified version of Algorithm 1 was considered to allow the cyclic execution of the discharge profile, i.e., as a duty cycle scheme, until the battery charge runs out. Through this simple modification, it becomes possible to predict the total battery lifetime according to both the discharge profile and the operating temperature. Scenarios 1 and 2 are used in these assessments as they depict the operating mode of traditional WSN nodes. The evaluations performed in this section consider the aspects below.

The first requirement to evaluate the battery lifetime estimation is to run the T-KiBaM model until the battery charge runs out. In these evaluations, the selected cutoff point occurs when the T-KiBaM algorithm indicates SoC = 0% ($\approx$2.0 V). It is worth mentioning that other cutoff points can be selected depending on the hardware requirements (e.g., 2.1 V or 2.2 V).

The second aspect concerns the tested set of tasks, which is the same as mentioned in Section 5.2 for Scenario 1. For simplification purposes, the experiments using Scenario 2 assume that the sleep mode does not consume energy (i.e., $I_2 = 0.0$ mA), although it is recognized that there is a small discharge current in this state, usually in the range of µA [54].

The last aspect concerns the number of iterations required for the algorithm to complete the estimation over the battery lifetime. In this case, the lower the granularity of the operating times (t_{I_x}) of the discharge currents (I_x), the greater the number of iterations of the algorithm and, consequently, the longer its computation time. Figure 5a depicts the number of iterations after performing the T-KiBaM model until the battery charge runs out for each discharge current presented in the mentioned set of tasks.

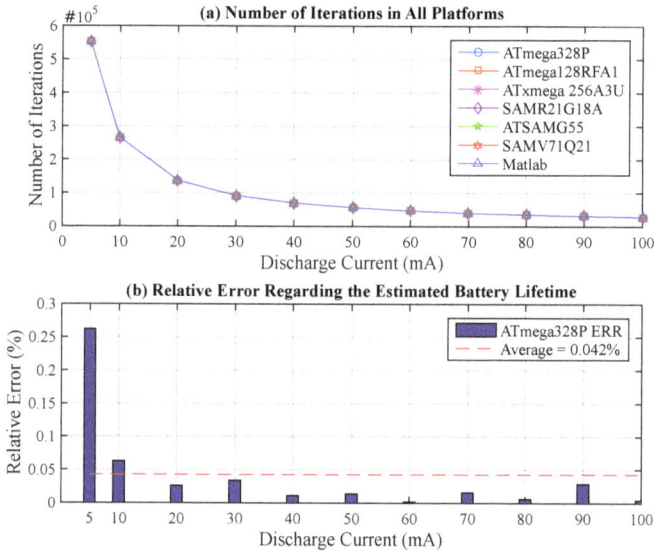

Figure 5. Results. (a) Number of iterations; (b) Relative Error (ERR).

Considering the previously mentioned aspects, the challenge of this evaluation is to assess how close the estimates of the battery lifetime are from the results obtained when executing the T-KiBaM model in a PC. The assessments for both Scenarios 1 and 2 are presented below.

For the Scenario 1 assessments, the entire set of tasks (i.e., $I_1 = 5, 10, 20, 30, 40, \ldots, 100$ mA) is executed with $t_1 = 1$ s. Table 9 presents the results regarding the estimated battery lifetime obtained when running the T-KiBaM algorithm on all platforms using Scenario 1. Note that the EXP column represents the results obtained when using real batteries at 25 °C, when available.

Table 9. Estimated battery lifetime *. ELT: Estimated Lifetime; EXP: Experimental time.

I_1 (mA)	ATmega 328P ELT (h)	ATmega128 RFA1 ELT (h)	ATxmega256 A3U ELT (h)	SAMR21 G18A ELT (h)	ATSAM G55 ELT (h)	SAMV71 Q21 ELT (h)	MATLAB ELT (h)	EXP (h)
5	153.1517	153.1517	153.1517	153.4969	153.4969	153.4969	153.5533	-
10	73.7003	73.7003	73.7003	73.6264	73.6264	73.6264	73.6536	73.557
20	37.8050	37.8050	37.8050	37.8028	37.8028	37.8028	37.8150	37.984
30	25.3956	25.3956	25.3956	25.3764	25.3764	25.3764	25.3869	25.385
40	19.1914	19.1914	19.1914	19.1850	19.1850	19.1850	19.1936	-
50	15.3572	15.3572	15.3572	15.3575	15.3575	15.3575	15.3550	-
60	12.7958	12.7958	12.7958	12.7947	12.7947	12.7947	12.7956	-
70	10.9658	10.9658	10.9658	10.9667	10.9667	10.9667	10.9675	-
80	9.5961	9.5961	9.5961	9.5953	9.5953	9.5953	9.5967	-
90	8.5328	8.5328	8.5328	8.5300	8.5300	8.5300	8.5303	-
100	7.6775	7.6775	7.6775	7.6778	7.6778	7.6778	7.6772	-

* Results considering Scenario 1 in all platforms.

The results indicate small relative Errors when compared to the estimated battery lifetime on a PC running MATLAB. For instance, considering all tested discharge currents, the average deviation between the ATmega328P and MATLAB is 0.042%. In this case, the minimum relative Error is 0.002% and the maximum relative Error is 0.262% (when $I_1 = 5$ mA). Figure 5b depicts the relative error of the ATmega328P with respect to the estimated battery lifetime when using T-KiBaM on MATLAB for the entire set of discharge currents. The other MCUs present the following average relative Errors: 0.042% (ATmega128RFA1), 0.042% (ATxmega256A3U), 0.023% (SAMR21G18A), 0.023% (ATSAMG55) and 0.023% (SAMV71Q21).

The evaluations for Scenario 2 consider the insertion of sleeping periods between the activities of the node, which operates in a duty cycle (DC) scheme. In this case, the evaluated duty cycles are as follows: 100%, 75%, 50%, 25%, 10%, and 5%. The discharge current (I_1) has its value set at 30 mA to allow comparison with the experimental results. Thus, the used current profiles are as follows:

$DC_{100\%} = [(30 \text{ mA}, 1 \text{ s}); (0.0 \text{ mA}, 0 \text{ s})]$, $DC_{75\%} = [(30.0 \text{ mA}, 3 \text{ s}); (0.0 \text{ mA}, 1 \text{ s})]$,

$DC_{50\%} = [(30.0 \text{ mA}, 1 \text{ s}); (0.0 \text{ mA}, 1 \text{ s})]$, $DC_{25\%} = [(30.0 \text{ mA}, 1 \text{ s}); (0.0 \text{ mA}, 3 \text{ s})]$,

$DC_{10\%} = [(30.0 \text{ mA}, 1 \text{ s}); (0.0 \text{ mA}, 9 \text{ s})]$, $DC_{5\%} = [(30.0 \text{ mA}, 1 \text{ s}); (0.0 \text{ mA}, 19 \text{ s})]$.

Since the results between the platforms for Scenario 1 are very close, the evaluations for Scenario 2 are performed only with the ATmega328P MCU. Table 10 presents the results obtained after running the T-KiBaM algorithm on this platform using Scenario 2. Again, the EXP column represents the results obtained from experiments with real batteries at 25 °C, when available.

Table 10. Estimated battery lifetime *. ELT: Estimated Lifetime; EXP: Experimental time.

Duty Cycle (%)	ATmega 328P ELT (h)	MATLAB ELT (h)	EXP (h)
100	25.3956	25.3869	25.385
75	33.8533	33.8489	33.524
50	50.7678	50.7744	51.229
25	101.6122	101.5489	102.547
10	253.8528	253.8722	-
5	507.3667	507.7444	-

* Results considering Scenario 2 in ATmega328P.

The results illustrated in Table 10 demonstrate that the estimates for the battery lifetime are compatible on both platforms. The variations in the results arise by virtue of the accuracy of the numerical representation in each platform. Regarding the voltage level tracking, Figure 6 depicts the behavior of the battery discharge curves for duty cycles of 75%, 50% and 25% at 25 °C. The experimental data represents the average behavior obtained in the experimental assessments, being presented as fitted curves. The analytical results are obtained through data prints during the execution of the T-KiBaM algorithm, however, only the fitted curves are presented for easy viewing.

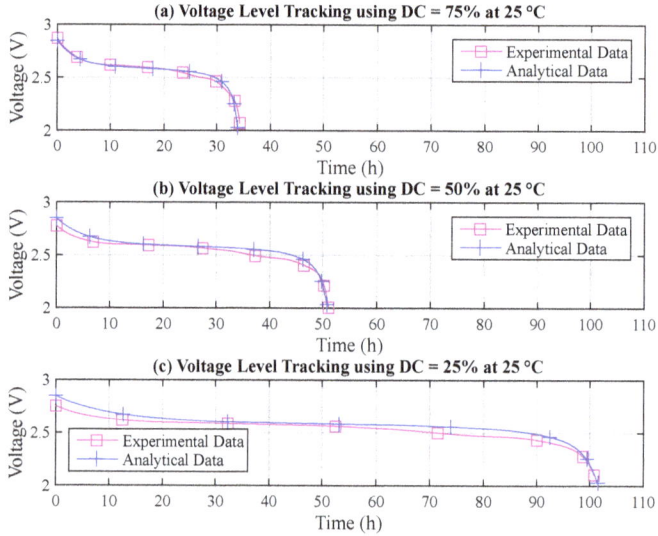

Figure 6. Results regarding the voltage level tracking. (**a**) DC = 75%; (**b**) DC = 50%; (**c**) DC = 25%.

6.3. Sensibility Analysis of T-KiBaM Model with Different Time Step Values

Finally, this section presents an assessment of the same application example, when different values are considered for the time step (t_{I_x}) of the discharge current (I_x) in the T-KiBaM function. The objective is to assess the relationship between the execution time of the algorithm for different tasks and the quality of the estimation prediction with respect to the battery operating behavior, i.e., its lifetime and voltage level over time, in Scenario 1. Note that the time step value corresponds to the interval between two consecutive invocations of the battery update function. The following time steps are used for this assessment: 1, 2, 5, 10 and 60 s. This assessment is performed only for the ATmega328P, as this micro-controller presents the hardware with the least amount of available resources among all the previously assessed devices. Thus, these results can be similarly extended to the other platforms.

First, the quality of the estimated battery lifetime is evaluated for different time steps. In this sense, the following metrics are evaluated: (i) execution time; (ii) number of iterations and (iii) estimated battery lifetime. The assessments considering Scenario 1 are performed below.

The first evaluated metric is the execution time for the entire set of tasks when different time steps are used as input to the T-KiBaM function. Figure 7a depicts the results obtained for the set of discharge currents mentioned in Section 5.2. Note that the execution time of each task (I_x, t_{I_x}) reduces dramatically, as the discharge current time step increases. For example, by comparing the time steps of 1 s and 10 s when $I_1 = 5$ mA, the execution time falls from 303.98 to 32.227 s when the algorithm is executed until the battery charge runs out. Considering the entire set of tasks, it is possible to observe an execution time 9.5 times faster, on average. The same behavior is observed for the second metric, i.e., the number of iterations, as shown in Figure 7b. Using the same time steps mentioned in the previous example, 1 s and 10 s, the number of iterations drops from 551,347 to 55,320, respectively.

Considering the entire set of tasks, it is possible to observe a reduction in the number of iterations equivalent to 10 times, on average.

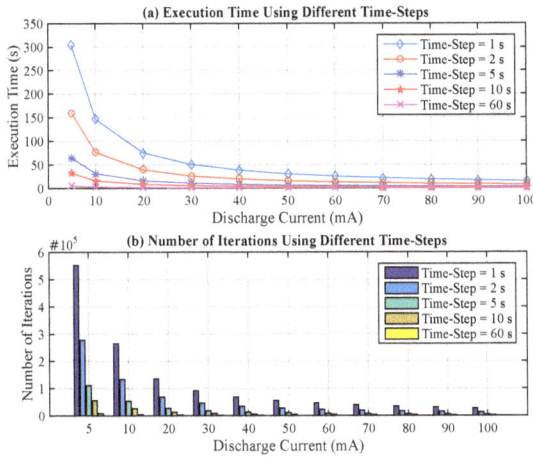

Figure 7. Results using different time steps. (**a**) Execution time; (**b**) Number of iterations.

The third evaluated metric is the estimated battery lifetime when different time steps are used as input to the T-KiBaM function. Figure 8a depicts the results obtained for the same set of discharge currents. As expected, the estimated battery lifetime has small variations for all the assessed cases.

Clearly, in terms of resource savings and performance, WSN designers should select the highest time step values. However, this selection must also take into account the imprecision introduced in the estimation when large time intervals are used to make the measurements. Figure 8b depicts the relative error for each time step considering the results obtained in a PC regarding the selected set of discharge currents. Note that the relative error is less than 0.4% for all the assessed cases. Particularly, the time step equal to 2 s has the highest relative errors for tasks with low discharge currents (<50 mA). On the other hand, the time step equal to 60 s presents highest relative error for tasks with larger discharge currents (>50 mA). According to these evaluations, the time step equal to 10 s is most indicated when continuous discharge currents are evaluated by the T-KiBaM model, since both the execution time and the number of iterations are significantly smaller and, at the same time, both the average relative error and the standard deviation compared to the values estimated in the PC are the lowest.

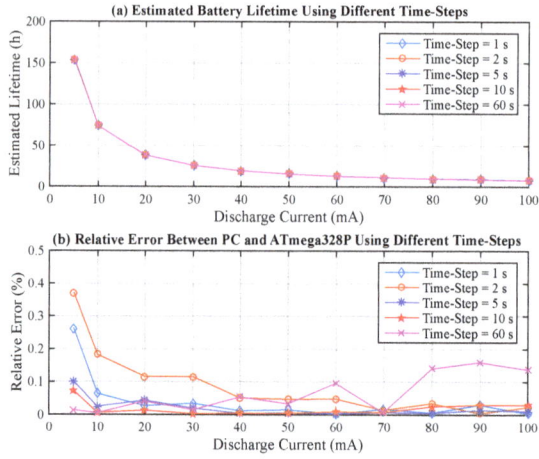

Figure 8. Results using different time steps. (**a**) Estimated battery lifetime; (**b**) Relative error.

Finally, the voltage level estimation provided by the T-KiBaM model is evaluated over time, using different time steps (Scenario 1). Again, a comparison of the experimental and analytical results is performed, using the results provided by the ATmega328P MCU at the time steps mentioned above. Figure 9 depicts the behavior of the estimation of the voltage curve at each update of the T-KiBaM model, under a continuous discharge current of 30 mA, at −5 °C. The experimental data are adjusted according to the average behavior of three experiments.

Note that the assessments done for the ATmega328P present the same results of the analytical evaluation performed on the PC, regardless the used time step. Thus, it is clear that the T-KiBaM model generates compatible results for both low-power and robust platforms regarding the voltage level tracking. This is a major result as estimating the voltage level over time is required to ensure the operation of any sensor node, allowing for optimizations in the WSN management policies.

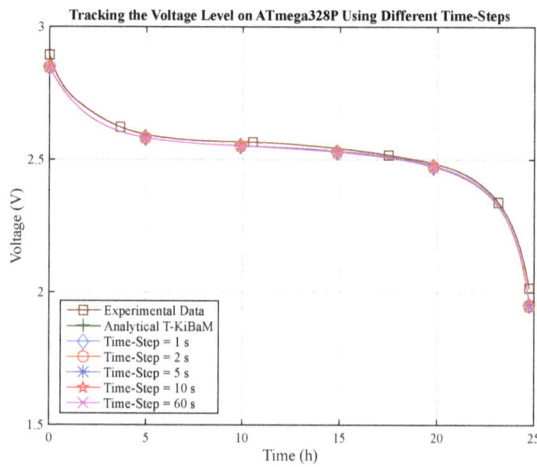

Figure 9. Results using different time steps for voltage tracking.

7. Conclusions

Estimating the battery lifetime is a complex task since many factors can influence the battery behavior, e.g., technology, operating temperature and discharge current. Analytical battery models may assist in this task, achieving results close to reality. However, two problems may arise within the WSN context. Firstly, the implementation of complex analytical models upon low-capacity hardware platforms is not an easy task, due to low processing capability, memory constraints and the high accuracy required to represent low varying analog values. Secondly, the execution of this type of analytical models by real-world nodes will influence its energy consumption, and therefore, the required effort to estimate the network lifetime may reduce the lifetime of the network itself.

The study performed in this paper evaluated the cost of executing an analytical battery model known as T-KiBaM in low-power MCUs. The model validation took into account experimental data. As shown in Section 4, the T-KiBaM model can accurately estimate the lifetime of Ni-MH batteries and is also able to estimate the voltage behavior over time at different temperatures, which is an important issue when considering devices (nodes) that require a minimum voltage value to maintain their operation. The analytical models were implemented upon different micro-controllers. As a result, although running T-KiBaM on low-power MCUs requires long computing times, such computing times do not represent a significant slice of the estimated battery lifetime. Therefore, the time required to estimate the battery behavior (which includes tracking both its SoC and voltage level over time) is feasible.

As future work, we are interested in finding a way to integrate the knowledge about both the discharge current and voltage level of the battery [36,55] to feed the T-KiBaM in real time, similar to the use of a fuel gauge IC in a smart battery pack. In this sense, it would become possible to implement a closed-loop approach, allowing the correction of predictions over time. Other issues will also be addressed in future work, such as a full research on the use of duty cycle discharge profiles at different temperatures. The validation of the estimates can also be made through the implementation of the proposed battery model in sensor nodes of a physical WSN. In this case, an application with a basic set of tasks should be used to allow the construction of a well-known fixed discharge profile. This would ensure minimal variability over the node's activities so that the results could be fairly comparable. Finally, the influence of the aging effect on sensor node batteries should be included in the proposed battery model to improve both the management and maintenance issues in WSNs.

Acknowledgments: The authors would like to thank the financial support from the CAPES/Brazil, CNPq/Brazil (Project 400508/2014-1; 445700/2014-9), FCT/Portugal (Project UID/EMS/50022/2013) and CAPES/FCT (Project 353/13) funding agencies.

Author Contributions: L.M.R., C.M., P.P. and F.V. conceived of and designed the experiments. P.P. and F.V. designed and built the test-bed for experimental assessments. L.M.R. conceived of T-KiBaM, performed the experiments and analytical evaluations and wrote the paper. G.B. tested the battery model and evaluated the power consumption by running the algorithm on the platforms mentioned in this paper. C.M., P.P. and F.V. provided guidance for writing and revised the paper.

Conflicts of Interest: The authors declare no conflict of interest.

J. Sens. Actuator Netw. **2017**, *6*, 8

Abbreviations

The following abbreviations are used in this manuscript:

ABI	Application Binary Interface
ARM	Advanced RISC Machine
CF	Correction Factor
COTS	Commercial Off-The-Shelf
DC	Duty Cycle
EEPROM	Electrically Erasable Programmable Read-Only Memory
ELT	Estimated Battery Lifetime
ET	Execution Time
ERR	Relative Error
FPU	Floating-Point Unit
IC	Integrated Circuit
KiBaM	Kinetic Battery Model
MAC	Media Access Control
MCU	Micro-Controller Unit
Ni-MH	Nickel-Metal Hydride
SoC	State of Charge
SRAM	Static Random Access Memory
T-KiBaM	Temperature-Dependent KiBaM
TVM	Temperature-Dependent Voltage Model
WSN	Wireless Sensor Network

References

1. Kim, T.; Qiao, W. A hybrid battery model capable of capturing dynamic circuit characteristics and nonlinear capacity effects. *IEEE Trans. Energy Convers.* **2011**, *26*, 1172–1180.
2. Wang, Y.; Zhang, C.; Chen, Z. A Method for state-of-charge estimation of LiFePO$_4$ batteries at dynamic currents and temperatures using particle filter. *J. Power Sources* **2015**, *279*, 306–311.
3. Lajara, R.J.; Perez-solano, J.J.; Pelegrí-sebastia, J. A method for modeling the battery state of charge in wireless sensor networks. *IEEE Sens. J.* **2015**, *15*, 1186–1197.
4. Gandolfo, D.; Brandão, A.; Patiño, D.; Molina, M. Dynamic model of lithium polymer battery–Load resistor method for electric parameters identification. *J. Energy Inst.* **2015**, *88*, 470–479.
5. Buchli, B.; Aschwanden, D.; Beutel, J. Battery state-of-charge approximation for energy harvesting embedded systems. In *Wireless Sensor Networks (EWSN)*; Demeester, P., Moerman, I., Terzis, A., Eds.; Springer: Berlin/Heidelberg, Germany, 2013.
6. Che, Z.; Jin, R.; Zhu, M.; Wang, Z.; Wang, L. Battery optimal scheduling based on energy balance in wireless sensor networks. *IET Wirel. Sens. Syst.* **2015**, *5*, 277–282.
7. Wang, Y.; Yang, D.; Zhang, X.; Chen, Z. Probability based remaining capacity estimation using data-driven and neural network model. *J. Power Sources* **2016**, *315*, 199–208.
8. Rahimi-Eichi, H.; Ojha, U.; Baronti, F.; Chow, M.Y. Battery management system: An overview of its application in the smart grid and electric vehicles. *IEEE Ind. Electron. Mag.* **2013**, *7*, 4–16.
9. Smart Battery System Implementers Forum. Smart Battery Data Specification–Addendum for Fuel Cell Systems, 2007. Available online: http://sbs-forum.org/specs/sbdata_addendum_fuel_cells_20070411.pdf (accessed on 31 March 2017).
10. Cadex Electronics Inc. Smart Battery Technology, 2017. Available online: http://www.cadex.com/en/batteries/smart-battery-technology (accessed on 31 March 2017).
11. Maxim Integrated. DS2780 Standalone Fuel Gauge IC. Available online: https://datasheets.maximintegrated.com/en/ds/DS2780.pdf (accessed on 18 April 2017).
12. Razaque, A.; Elleithy, K. Energy-efficient boarder node medium access control protocol for wireless sensor networks. *Sensors* **2014**, *14*, 5074–5117.
13. Jabbar, S.; Minhas, A.A.; Imran, M.; Khalid, S.; Saleem, K. Energy efficient strategy for throughput improvement in wireless sensor networks. *Sensors* **2015**, *15*, 2473–2495.
14. Mammu, A.S.I.K.; Hernandez-Jayo, U.; Sainz, N.; de la Iglesia, I. Cross-layer cluster-based energy-efficient protocol for wireless sensor networks. *Sensors* **2015**, *15*, 8314–8336.

15. Jongerden, M.R.; Haverkort, B.R. *Battery Modeling;* Technical Report; University of Twente: Enschede, The Netherlands, 2008.
16. Rodrigues, L.M.; Montez, C.; Moraes, R.; Portugal, P.; Vasques, F. A temperature-dependent battery model for wireless sensor networks. *Sensors* **2017**, *17*, 422.
17. Atmel Corporation. Home Page. Available online: http://www.atmel.com (accessed on 2 June 2017).
18. Cheng, P.; Zhou, Y.; Song, Z.; Ou, Y. Modeling and SOC estimation of LiFePO$_4$ battery. In Proceedings of the IEEE International Conference on Robotics and Biomimetics (ROBIO), Qingdao, China, 3–7 December 2016.
19. Wang, Y.; Zhang, C.; Chen, Z. A method for state-of-charge estimation of Li-Ion batteries based on multi-model switching strategy. *Appl. Energy* **2015**, *137*, 427–434.
20. Zhu, Q.; Xiong, N.; Yang, M.L.; Huang, R.S.; Hu, G.D. State of charge estimation for Lithium-Ion battery on nonlinear observer: An H_{inf} method. *Energies* **2017**, *10*, 1–19.
21. Hannan, M.A.; Lipu, M.S.H.; Hussain, A.; Mohamed, A. A review of Lithium-Ion battery state of charge estimation and management system in electric vehicle applications: Challenges and recommendations. *Renew. Sustain. Energy Rev.* **2017**, *78*, 834–854.
22. Çakiroğlu, M. Software implementation and performance comparison of popular block ciphers on 8-bit low-cost microcontroller. *Int. J. Phys. Sci.* **2010**, *5*, 1338–1343.
23. Liu, W.; Luo, R.; Yang, H. Cryptography overhead evaluation and analysis for wireless sensor networks. In Proceedings of the WRI International Conference on Communications and Mobile Computing, Yunnan, China, 6–8 January 2009; pp. 496–501.
24. Capo-Chichi, E.P.; Guyennet, H.; Friedt, J.M. K-RLE: A new data compression algorithm for wireless sensor networks. In Proceedings of the International Conference on Sensor Technologies and Applications (SENSORCOMM), Athens, Greece, 18–23 June 2009; pp. 502–507.
25. Guo, H.; Low, K.S.; Nguyen, H.A. Optimizing the localization of a wireless sensor network in real time based on a low-cost microcontroller. *IEEE Trans. Ind. Electron.* **2011**, *58*, 741–749.
26. Othman, S.B. Performance evaluation of encryption algorithm for wireless sensor networks. In Proceedings of the International Conference on Information Technology and e-Services (ICITeS), Sousse, Tunisia, 24–26 March 2012; pp. 1–8.
27. Quirino, G.S.; Moreno, E.D.; Matos, L.B.C. Performance evaluation of asymmetric encryption algorithms in embedded platforms used in WSN. In Proceedings of the World Congress in Computer Science, Computer Engineering and Applied Computing (WORLDCOMP), Las Vegas, NV, USA, 22–25 July 2013.
28. Pardo, J.; Zamora-Martínez, F.; Botella-Rocamora, P. Online learning algorithm for time series forecasting suitable for low cost wireless sensor networks nodes. *Sensors* **2015**, *15*, 9277–9304.
29. Panić, G.; Stecklina, O.; Stamenković, Z. An embedded sensor node microcontroller with crypto-processors. *Sensors* **2016**, *16*, 607.
30. Leveque, A.; Pecheux, F.; Louerat, M.M.; Aboushady, H.; Vasilevski, M. SystemC-AMS models for low-power heterogeneous designs: Application to a WSN for the detection of seismic perturbations. In Proceedings of the International Conference on Architecture of Computing Systems (ARCS), Hannover, Germany, 22–25 February; pp. 1–6.
31. Biswas, K.; Muthukkumarasamy, V.; Wu, X.W.; Singh, K. An analytical model for lifetime estimation of wireless sensor networks. *IEEE Commun. Lett.* **2015**, *19*, 1584–1587.
32. Kim, J.U.; Kang, M.J.; Yi, J.M.; Noh, D.K. A simple but accurate estimation of residual energy for reliable WSN applications. *Int. J. Distrib. Sens. Netw.* **2015**, *2015*, doi:10.1155/2015/107627.
33. Dron, W.; Duquennoy, S.; Voigt, T.; Hachicha, K.; Garda, P. An emulation-based method for lifetime estimation of wireless sensor networks. In Proceedings of the IEEE International Conference on Distributed Computing in Sensor Systems, Marina Del Ray, CA, USA, 25–27 May 2014; pp. 241–248.
34. Rahmé, J.; Fourty, N.; Al Agha, K.; Van Den Bossche, A. A recursive battery model for nodes lifetime estimation in wireless sensor networks. In Proceedings of the IEEE Wireless Communications and Networking Conference (WCNC), Sydney, Australia, 18–21 April 2010.
35. Rakhmatov, D.; Vrudhula, S. Energy management for battery-powered embedded systems. *ACM Trans. Embed. Comput. Syst.* **2003**, *2*, 277–324.
36. Kerasiotis, F.; Prayati, A.; Antonopoulos, C.; Koulamas, C.; Papadopoulos, G. Battery lifetime prediction model for a WSN platform. In Proceedings of the International Conference on Sensor Technologies and Applications (SENSORCOMM), Venice, Italy, 18–25 July 2010; pp. 525–530.

37. Nataf, E.; Festor, O. Online estimation of battery lifetime for wireless sensor network. *arXiv* **2012**, arXiv:1209.2234.

38. Rukpakavong, W.; Guan, L.; Phillips, I. Dynamic node lifetime estimation for wireless sensor networks. *IEEE Sens. J.* **2014**, *14*, 1370–1379.

39. Park, C.; Lahiri, K.; Raghunathan, A. Battery discharge characteristics of wireless sensor nodes: An experimental analysis. In Procedings of the Sensor and Ad Hoc Communications and Networks, Santa Clara, CA, USA, 26–29 September 2005; pp. 430–440.

40. Manwell, J.F.; McGowan, J.G. Lead acid battery storage model for hybrid energy systems. *Solar Energy* **1993**, *50*, 399–405.

41. Manwell, J.F.; McGowan, J.G. Extension of the kinetic battery model for wind/hybrid power systems. In Proceedings of the European Wind Energy Association Conference (EWEC), Thessaloniki, Greece, 10–14 October 1994; pp. 284–289.

42. Chen, M.; Rincon-Mora, G. Accurate electrical battery model capable of predicting runtime and I–V performance. *IEEE Trans. Energy Convers.* **2006**, *21*, 504–511.

43. Jaguemont, J.; Boulon, L.; Venet, P.; Dube, Y.; Sari, A. Lithium-Ion battery aging experiments at subzero temperatures and model development for capacity fade estimation. *IEEE Trans. Veh. Technol.* **2016**, *65*, 4328–4343.

44. Tremblay, O.; Dessaint, L.A.; Dekkiche, A.I. A generic battery model for the dynamic simulation of hybrid electric vehicles. In Procedings of the Vehicle Power and Propulsion Conference (VPPC), Arlington, TX, USA, 9–12 September 2007; pp. 284–289.

45. Tremblay, O.; Dessaint, L.A. Experimental validation of a battery dynamic model for EV applications. *World Electr. Veh. J.* **2009**, *3*, 1–10.

46. Atmel Corporation. 8-Bit AVR Microcontroller with 4/8/16/32K Bytes In-System Programmable Flash, 2017. Available online: http://www.atmel.com/pt/br/devices/ATMEGA328P.aspx (accessed on 9 February 2017).

47. Atmel Corporation. 8-Bit AVR Microcontroller with Low Power 2.4 GHz Transceiver for ZigBee and IEEE 802.15.4 (ATmega128RFA1), 2017. Available online: http://www.atmel.com/pt/br/Images/Atmel-8266-MCU_Wireless-ATmega128RFA1_Summary_Datasheet.pdf (accessed on 9 February 2017).

48. Atmel Corporation. 8/16-bit Atmel XMEGA A3U Microcontroller, 2017. Available online: http://www.microchip.com/wwwproducts/en/ATxmega256A3U (accessed on 6 March 2017).

49. Atmel Corporation. SMART ARM-Based Wireless Microcontroller, 2017. Available online: http://www.atmel.com/Images/Atmel-42223\T1\textendashSAM-R21_Datasheet.pdf (accessed on 6 March 2017).

50. Atmel Corporation. ATSAMG55, 2017. Available online: http://ww1.microchip.com/downloads/en/DeviceDoc/Atmel-11289-32-bit-Cortex-M4-Microcontroller-SAM-G55_Datasheet.pdf (accessed on 9 March 2017).

51. Atmel Corporation. ATSAMV, 2017. Available online: http://www.atmel.com/products/microcontrollers/arm/sam-v-mcus.aspx (accessed on 3 June 2017).

52. launchpad.net. Poorly Optimised Code Generation for Cortex M0/M0+/M1 vs M3/M4, 2017. Available online: https://bugs.launchpad.net/gcc-arm-embedded/+bug/1502611 (accessed on 2 June 2017).

53. Liu, X.; Hou, K.M.; Vaulx, C.D.; Shi, H.; Gholami, K.E. MIROS: A hybrid real-time energy-efficient operating system for the resource-constrained wireless sensor nodes. *Sensors* **2014**, *14*, 17621–17654.

54. Mikhaylov, K.; Tervonen, J. Node's Power Source Type Identification in Wireless Sensor Networks. In Proceedings of the International Conference on Broadband and Wireless Computing, Communication and Applications, Barcelona, Spain, 26–28 October 2011; pp. 521–525.

55. Barboni, L.; Valle, M. Experimental analysis of wireless sensor nodes current consumption. In Proceedings of the International Conference on Sensor Technologies and Applications (SENSORCOMM), Cap Esterel, France, 25–31 August 2008; pp. 401–406.

Journal of
*Sensor and
Actuator Networks*

MDPI

Article

An SVM-Based Method for Classification of External Interference in Industrial Wireless Sensor and Actuator Networks

Simone Grimaldi *, Aamir Mahmood and Mikael Gidlund

Department of Information Systems and Technology, Mid Sweden University, 851 70 Sundsvall, Sweden;
aamir.mahmood@miun.se (A.M.); mikael.gidlund@miun.se (M.G.)
* Correspondence: simone.grimaldi@miun.se; Tel.: +46-010-142-8249

Academic Editor: Mário Alves
Received: 30 April 2017; Accepted: 12 June 2017; Published: 16 June 2017

Abstract: In recent years, the adoption of industrial wireless sensor and actuator networks (IWSANs) has greatly increased. However, the time-critical performance of IWSANs is considerably affected by external sources of interference. In particular, when an IEEE 802.11 network is coexisting in the same environment, a significant drop in communication reliability is observed. This, in turn, represents one of the main challenges for a wide-scale adoption of IWSAN. Interference classification through spectrum sensing is a possible step towards interference mitigation, but the long sampling window required by many of the approaches in the literature undermines their run-time applicability in time-slotted channel hopping (TSCH)-based IWSAN. Aiming at minimizing both the sensing time and the memory footprint of the collected samples, a centralized interference classifier based on support vector machines (SVMs) is introduced in this article. The proposed mechanism, tested with sample traces collected in industrial scenarios, enables the classification of interference from IEEE 802.11 networks and microwave ovens, while ensuring high classification accuracy with a sensing duration below 300 ms. In addition, the obtained results show that the fast classification together with a contained sampling frequency ensure the suitability of the method for TSCH-based IWSAN.

Keywords: industrial wireless sensor and actuator networks; support vector machine; interference classification; spectrum-sensing; wireless LAN; microwave oven

---

1. Introduction

The use of wireless sensor networks (WSNs) is a growing trend in a myriad of application domains, including building-health monitoring [1], military applications [2], health monitoring systems [3] and disaster and emergency management [4], to mention a few. A common denominator for many of these networks is the underlying radio technology, which is based on the IEEE 802.15.4 standard [5]. However, depending on the application, different requirements are set regarding the quality of service (QoS). In particular, differently from common implementations of WSN, the requirements found in those deployed in industrial settings, also known as industrial wireless and actuator networks (IWSANs), are considerably more challenging. Furthermore, the inclusion of actuators allows the IWSAN to cover more specific applications, such as closed-loop control, in which bi-directional data-traffic is needed.

IWSANs are characterized by having star or few hops mesh topology with a small number of devices and for presenting stringent requirements on the end-to-end communication delay and reliability. These requirements commonly include downlink and uplink transmission of process data with refresh rates in the order of tens of milliseconds and a network uptime greater than 99.999%, which corresponds to a downtime of less than 5.26 min per year [6]. Fulfilling such communication requirements is critically important in order to enable the adoption of IWSAN as a replacement of traditional wired implementations, such as Fieldbus-based solutions [7]. A failure to meet the QoS requirements can result in unwanted and costly production halts, corruption of the industrial product or even physical damage to production devices and human harm.

The two main factors that hamper the performance of IWSANs are the harsh radio-propagation conditions of most industrial environments, with pronounced effects of multipath fading and attenuation (MFA), and the interference originated from RF emissions in the 2.4-GHz unlicensed industrial, scientific, and medical (ISM)-band. The combined effect of these phenomena can cause severe degradation of the IWSAN radio links, potentially generating prolonged communication outages in some sectors of the wireless network. The RF interference that affects IEEE 802.15.4-based WSNs is mainly generated by wireless systems sharing the same ISM-band and microwave ovens (MWO), while the RF emissions of other devices (e.g., electric motors or switches) is mainly confined to the sub-GHz region of the spectrum, as shown in [8] and the references therein. Nevertheless, while some industrial plants can employ MWO in their production process (e.g., industrial material drying or food processing [9]), the wireless systems that reside in the 2.4-GHz band are much more frequent. The most widespread technologies that operate in this band are the IEEE 802.11 and IEEE 802.15.1 standards, under the commercial name of Wi-Fi and Bluetooth, respectively. IEEE 802.11-based WLANs are generally acknowledged as the most severe cause of interference for a number of reasons. Primarily, IEEE 802.11 networks are now ubiquitous in both office and production areas due to the widespread diffusion of WiFi-enabled terminals, such as smartphones or laptops. Moreover, in order to achieve full coverage, numerous access points are deployed, which can represent an obstacle for coexistence with IWSANs. Additionally, the IEEE 802.11 standard defines a physical layer (PHY), which enables transmission powers ten-times higher than IEEE 802.15.4 devices and a 5–8-times wider channel bandwidth, as shown in Figure 1. As a result, a coexisting IEEE 802.11 network can cause a packet error rate (PER) up to 70% [10–12] for a WSN receiver under the worst-case scenarios, such as prolonged use of overlapping channels, proximity of an IEEE 802.11 access point and sustained utilization rate of the interfering network. While devices implementing the IEEE 802.15.1 standard can also be found in industrial settings, thanks to the limited channel bandwidth and the implemented frequency-hopping scheme, their impact on the performance of IEEE 802.15.4-based networks is limited compared to MWO and IEEE 802.11 interference, as reported in [13]. For this reason, the classification of IEEE 802.15.1 interference is not considered in this paper.

Time-slotted channel-hopping (TSCH) is a well-known technique implemented in IWSAN standards, including WirelessHART [14], ISA100.11a [15] and WIA-PA [16], to mitigate the effects of external interference. Nevertheless, none of these standards employs intelligent methods for classifying the source of interference and adopting ad hoc strategies for interference mitigation. Since the first release of the IEEE 802.15.4 standard in 2003, a consistent number of research works has been carried out addressing interference-awareness in WSN. This matter can be separated into two different, but tightly-related aspects: interference classification and interference mitigation. In the terminology of cognitive-radio systems [17], the secondary-users (i.e., WSN-devices) are required to gain a certain level of spectrum awareness in order to utilize the unused resources opportunistically. A common approach for spectrum sensing methods in the literature is to adopt a relatively high sampling frequency and a sensing-time in the order of seconds, in order to maximize classification accuracy or make inference on the inflicted PER [18]. However, this is not suitable in the context of the time-critical IWSANs, where a long spectrum-sensing time implies slow network reactivity to the variations of the interference-scenario and waste of network resources due to the need for reserving numerous silent timeslots for channel sensing.

In this article, an interference detection and classification method is proposed and analyzed, with particular focus on minimizing the time required for channel sensing and the complexity of feature selection, while ensuring a good level of in-channel detection accuracy. For this purpose, a distributed spectrum sensing strategy and a centralized classification algorithm are employed to generate a space-frequency map of interference-free channels (IFCs). The IFC map is valuable information in the context of interference-aware resource scheduling for interference mitigation. The proposed interference classifier uses a three-step classification strategy, comprising a lightweight feature extraction stage, a set of four support vector machines (SVMs) performing preliminary binary classifications and a final stage composed of a logic decisor. The introduced mechanism is able to discriminate among interference from IEEE 802.11 networks, even when no terminal is associated with the access point, RF leakage from MWO and an IFC. Differently from other methods in the literature, such as [19–22], the proposed method does not rely on features based on the periodicity of IEEE 802.11 beacons. This fact, in conjunction with the novel classification scheme based on multiple SVMs, helps to ensure good classification performance while requiring an extremely limited sensing time.

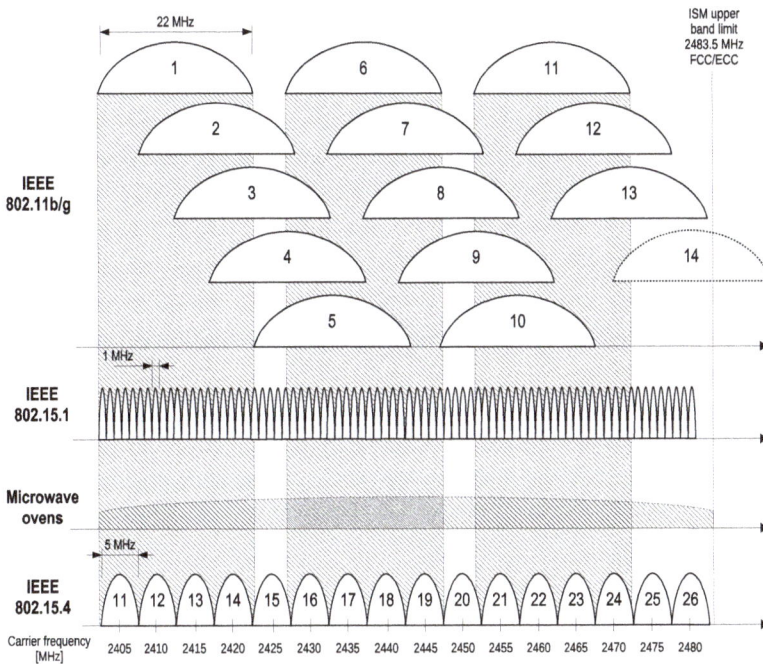

Figure 1. The 2.4-GHz industrial, scientific and medical (ISM) spectrum. Channel allocation for heterogeneous technologies with RF emissions within the band: IEEE 802.11, Bluetooth, microwave oven (MWO), IEEE 802.15.4.

The main contributions of this work are as follows:

- This is the first study that employs an SVM classifier to process signal features extracted from received signal strength indicator (RSSI) traces to identify the source of external interference. The proposed method employs four lightweight signal features, designed considering hardware constraints of commercial off-the-shelf (COTS) WSN devices.
- It is shown that, in order to ensure good detection performance, the proposed classifier requires a time window for spectrum sensing consistently below 300 ms, which, to the best knowledge of the authors, places the proposed solution amongst the quickest and most reliable methods reported in the literature.
- The performance of the proposed solution is validated by using an RSSI dataset collected in different industrial environments. Both the controlled and uncontrolled interferences from IEEE 802.11 networks are taken into account.

- The often overlooked influence of device calibration on spectrum sensing-based interference classification is analyzed, showing that the classifier accuracy is subject to the intrinsic hardware variations of the employed devices. However, we show that this factor can be easily corrected by means of a straightforward calibration process.

The remainder of this article is structured as follows. In Section 2, relevant work available in the literature about interference classification and mitigation in WSNs is presented. Section 3 provides a general background of the topic, discussing the various sources of cross-technology interference, with specific interest in the IEEE 802.11 standard. In Section 4, the basic concepts and mathematical formulation for SVMs are explained. In Section 5, a detailed description of the proposed solution is given, highlighting feature selection and the structure of the proposed classifier. In Section 6 and Section 7, the experimental setup and the results from experiments are described. In Section 8, the achieved results are discussed, and lastly, conclusions and final considerations are drawn in Section 9.

2. Related Works

The unrestricted and widespread usage of the unlicensed 2.4-GHz ISM bands, coupled with the asymmetric transmit power and medium access rules, results in harmful mutual interference among coexisting wireless systems. The most affected are the low-power systems, such as IEEE 802.15.4-based WSNs. Various experimental and theoretical studies have highlighted WSNs' susceptibility to the external interference, especially from high transmit power IEEE 802.11-based WLANs. Many experimental studies (e.g., [10,23,24]) show that an IEEE 802.15.4 link operating on a channel overlapped by an IEEE 802.11 network can experience packet losses of up to 50–70%. In light of these performance studies, it is evident that without an interference detection and avoidance mechanism, WSNs cannot satisfy any reliability or dependability conditions required by the aforementioned industrial applications. The most common interference detection technique, also recommended by the ZigBee standard [25], is to utilize energy detection-based spectrum sensing and avoid the channels with an energy level above a certain threshold. However, in order to design an intelligent interference avoidance technique, the type of interference and its behavior in the time and frequency domains need to be identified first. As interference scenarios may evolve in time, adaptive mitigation approaches with an individual strategy are efficient and are recommended [11]. There exist two main approaches to interference classification, where the distinction is made based on the information source used to extract the features to analyze: (i) raw channel energy measurements (i.e., RSSI samples), and (ii) bit error patterns in a corrupted packet. The existing methods for interference classification available in the literature are shown in Figure 2 and further discussed in the following sections.

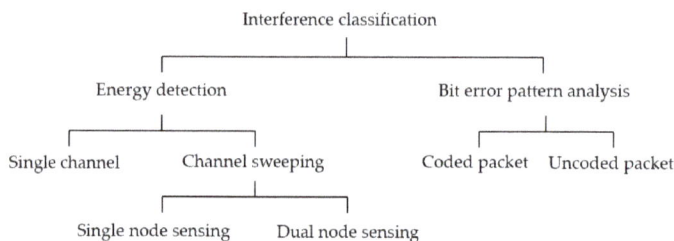

Figure 2. An overview of interference classification methods.

2.1. Energy Detection-Based Interference Classification

In this approach, a node actively collects energy samples on one or more IEEE 802.15.4 channels when the WSN devices are not transmitting. Signal processing techniques are then applied to the stored samples in order to extract a number of signal features, according to the implemented method.

Hard conditioning or machine learning techniques are then employed in order to map these features to a class of interference, such as IEEE 802.11, Bluetooth and MWO. The advantage of this approach is that no packet transmission is required, since there is no feature extraction from received packets. On the other hand, these methods require a certain time window in order to collect the required energy samples, meaning that specific idle-periods have to be reserved for channel sensing, potentially reducing the availability of network resources for data transmission.

In [19,26], Zacharias et al. propose a lightweight interference classification method in which a series of conditions are tested to identify the dominant source of interference. In these works, a node collects the RSSI samples on a single channel over a duration of one second at a sampling frequency of 8 kHz. The samples are then binarized using a fixed threshold of −85 dBm. Based on the binary data, the temporal features such as channel idle, busy time and signal periodicity are extracted. The classification conditions are then applied to the extracted features to identify the type of interference, achieving a classification time between 600 ms and 700 ms.

The detection of multiple sources of interference is studied by the authors in [27]. In this study, a clustering algorithm is applied to RSSI samples (collected by a node at a sampling frequency of 21 kHz) to distinguish the RSSI bursts from different interferers. In addition, a classifier identifies the channel activity patterns as periodic, bursty or a combination of both to determine channel suitability. The identification of periodic signals such as IEEE 802.11 beacons is also considered, achieving a classification accuracy of over 90% for sampling windows greater than or equal to 3 s.

The detection of the IEEE 802.11 beacons for the discovery of an IEEE 802.11 network has also been the subject of investigation in [20–22]. The collection of RSSI measurements over available channel sets (channel sweeping) is considered in [28], which employs two IEEE 802.15.4 radios to achieve pair-wise synchronized channel sensing. The objective of collecting samples over multiple channels is to identify the interference by matching the observed spectral pattern with the stored reference shape. The work targets only IEEE 802.11 interference, achieving a classification rate of 96% with sensing time in the order of 300 ms. In [29], instead, a single node is used for channel sweeping, and an interference classification method targeting IEEE 802.11 and MWOs is proposed in the context of an interferer-aware transmission adaptation mechanism.

Based on the high-resolution scanning feature of Atheros-based WLAN cards, the authors in [30] were able to extract detailed timing and frequency information of the interfering signal, at the cost of using additional hardware. In this context, a decision-tree classifier for interference identification was implemented yielding 91–96% detection accuracy. In [31], Weng et al. have developed two algorithms for the identification of MWO, IEEE 802.11 and Bluetooth signals based on 20 MHz I/Q data sampling performed by means of additional spectrum sensing hardware. However, these approaches are beyond the scope of the hardware capability of commonly-used resource-constrained sensor nodes.

2.2. Bit Error Pattern-Based Interference Classification

This class of methods does not require the active collection of RSSI samples; rather, the interference classification is based on the analysis of bit error patterns in the packets exchanged during the normal operation of the network. In [32,33], the authors show that the different interferers, such as IEEE 802.11, Bluetooth and ZigBee, corrupt IEEE 802.15.4 packets, leaving specific error footprints. In addition, the bit error pattern can also be used to reveal the presence of weak links. In particular, Hermans et al. [32] propose identification of the interference source by combining (i) the signal strength variations during packet reception, (ii) the link quality indicator (LQI) associated with a packet and (iii) the position of corrupted bytes in the payload. The classification accuracy of the proposed method is 72%, while this result is also IEEE 802.15.4 packet size dependent, since packets with a small payload size are partially overlapped with the interferer, thus carrying a small interference fingerprint. In [34] Barac et al. use forward error correction (FEC) in order to identify the source of bit errors in a received packet (i.e., multipath fading and attenuation, as well as the IEEE 802.11 b/g interference). Therefore, instead of packet retransmissions (which is used in [32]), the FEC method in [34] emerges as an energy-efficient alternative for interference classification, yielding more than 91% classification rate with just one received packet.

3. Background

3.1. Cross-Technology Interference Sources

In this section, we discuss the salient features of the cross-technology sources of interference targeted in the current work, namely IEEE 802.11-based WLAN and MWO.

3.1.1. IEEE 802.11

The prevalent WLAN networks in the 2.4-GHz band are based on the IEEE 802.11 b/g/n specifications. The IEEE 802.11 b/g PHY supports up to 14 channels, 20 MHz wide each. On the other hand, the IEEE 802.11 n can support both the 20 MHz- or 40 MHz-wide channels. There are only three non-overlapping usable channels in the U.S. and other countries with similar regulations (Channels 1, 6, 11, with 25-MHz separation) and four in Europe (Channels 1, 5, 9, 13, with 20-MHz separation). The transmit power of WLAN devices ranges from 15 dBm–20 dBm, and depending on the underlying standard, different modulation schemes and data-rates are available. However, the maximum air-time of an IEEE 802.11 packet remains below 600 μs. The standard specifies a carrier-sense multiple access with collision-avoidance (CSMA/CA)-based MAC with certain timing rules between the consecutive packets. In commonly-used infrastructure mode, an access point advertises the network by sending the periodic beacon frames. For compatibility reasons and in order to increase the network detection range, the beacons are usually sent at the lowest data rate (1 or 2 Mb/s). The default beacon frequency period is 100 time units, which is equal to 102.4 ms [26]. The above-stated heterogeneous medium access rules and PHY specifications for WLAN networks render ZigBee systems vulnerable. Firstly, WLAN networks deployed on non-overlapping channel allocation, such as the typical {1, 6, 11} configuration, leave a small number of IFCs for ZigBee. Secondly, the high-power concurrent transmission on an overlapping channel from a WLAN device will likely cause severe packet corruption in an IEEE 802.15.4 packet. Thirdly, the duration of an IEEE 802.15.4 clear-channel assessment (CCA) is 128 μs, while 192 μs [14] are required to switch from CCA to transmit mode. Conversely, the IEEE 802.11 CCA procedure takes 28 μs, while the switching time is negligible. As a result, the chances of the corruption of a ZigBee packet are very high, as the WLAN transmission can disrupt the ZigBee transmission during the switching mode.

3.1.2. Microwave Ovens

The energy leakage from the residential MWOs usually affects the whole 2.4-GHz band. However, as depicted by various studies [29,35], the RF emissions from MWOs peak at about a 2.45-GHz frequency, while the number and center frequencies of peaks may vary slightly according to the specific model, as shown in [36]. As a result, the IEEE 802.15.4 Channels 20 and 21 have a high probability of being strongly affected by the MWO operation. A prominent feature of MWO is the periodicity of *on* and *off* phases during the heating process, where the time from one *on* phase to the next is $\frac{1}{2f}$ s, with f the frequency of the power supply (i.e., 50/60 Hz).

4. Support Vector Machines

In this section, we outline the basic formulation of the mathematical problem for an SVM, focusing on the training and classification tasks, while we leave more in-depth analysis to specific machine-learning literature, such as [37], and to [38] for details about the related convex optimization methods.

4.1. The Standard Model for SVM

An SVM is a supervised classification algorithm that allows a binary decision to be performed, assigning an M-dimensional feature-vector to one of two classes. Being a supervised approach, an SVM needs to be trained using an appropriate dataset, which should be sufficiently large and representative of the two classes, with respect to the selected features. A training phase is then needed to determine a subset of the training vectors (called support vectors), which will actually be used for solving the classification problem. One important advantage of SVMs resides in the fact that the number of support vectors is generally much smaller than the cardinality of the training dataset. Hence, while the training of the SVM can be a resource-intensive task, the actual classification algorithm can be very slender. The standard formulation for a two-class classification problem is:

$$y(\mathbf{x}) = \mathbf{w}^T \phi(\mathbf{x}) + b \tag{1}$$

which is a linear model where $\mathbf{x}$ is the M-dimensional input vector, M is the size of the feature space, $\mathbf{w} = \{w_1, w_2, \dots, w_M\}$ is the vector of coefficients for the linear model, ϕ is a general feature-space transformation function (which can eventually be non-linear) and b represents the bias of the model.

Hence, the training set for the SVM is composed of a set of N training feature-vectors $\mathbf{x}_1, \dots, \mathbf{x}_N$ where each vector is associated with one of the two classes (C_1, C_2) via the parameters $t_n = \{-1, 1\}$, which are the class labels for the training vectors. The decision logic is then the following: an unknown vector $\mathbf{x}^*$ belongs to class C_1, if $y(\mathbf{x}^*) < 0$ and to class C_2 if $y(\mathbf{x}^*) > 0$. The implicit assumption is that the training data are linearly separable, so that the coefficient vector $\mathbf{w}$ and the parameter b can be determined (i.e., there exists at least one feasible combination of $\mathbf{w}$ and b).

4.2. SVM: Training and Classification

The training of an SVM can be seen geometrically as the problem of maximizing the minimum Euclidean distance between the decision hyperplane and the points of the training set. This problem can be formulated in an equivalent fashion, observing that since $t_n = \{-1, 1\}$ are the target values for the two classes, the following is verified for any correctly-labeled input vector $\mathbf{x}$:

$$t_n y(\mathbf{x}) > 0 \tag{2}$$

It can be easily shown that the optimization problem can be expressed as:

$$\begin{aligned} \text{minimize} \quad & \|\mathbf{w}\|_2 \\ \text{subject to:} \quad & t_n(\mathbf{w}^T \phi(\mathbf{x}_n) + b) - 1 > 0 \end{aligned} \tag{3}$$

with $n \in [1, N]$ Hence, due to the definition of two norm, the function to minimize in (3) is a quadratic cost function with M variables. The optimization problem that arises is then a quadratic program (QP) with M variables (size of the feature space) and N inequality constraints (size of the RSSI input vector).

Once the model is trained, the solution of the decision problem for a generic input vector $\mathbf{x}^*$ can be obtained by simply evaluating the sign of $y(\mathbf{x})$ in the original linear model $y(\mathbf{x}) = \mathbf{w}^T \phi(\mathbf{x}) + b$, with the coefficient vector $\mathbf{w}$ populated using the results from the minimization of the cost function in (3), hence calculating:

$$y(\mathbf{x}) = \sum_{n=1}^{N} t_n \alpha_n k(x, x_n) \tag{4}$$

where α_n are the Lagrange multipliers of the dual problem. Equation (4) is subject to the Karush-Kuhn-Tucker (KKT) conditions:

$$\alpha_n \geq 0$$

$$t_n y(x_n) \geq 0$$

$$\alpha_n (t_n y(x_n) - 1) = 0$$

An important result is that each point of the cost function for which the respective Lagrange multiplier $\alpha_n = 0$ can be discharged, since it will not influence the calculation, yields a consistent reduction of the dataset size, which is one of the key advantages of SVMs.

5. The Proposed Solution

5.1. Classifier Setup

The proposed interference detection method employs an SVM-based classifier, which processes input data composed of observations of the background RF noise on a specific IEEE 802.15.4 radio channel. The method is based on the basic assumption that when there is no transmission on a certain channel (and thus, there is an absence of intra-network interference), the devices can collect samples of the RF radiation and process the data to detect and classify eventual interferers, as well as assessing the eventuality of an IFC. This assumption nicely fits with the time-division multiple-access (TDMA) approach employed in ISWANs, since in these networks, the allocation of frequency-time resources for data transmission is known a priori; thus, a contiguous set of time slots on a specific channel can be reserved for spectrum sensing. The common hypothesis for spectrum sensing is that the classification can be done with a certain level of accuracy if the time window is sufficiently long for specific signal features to emerge. The proposed solution is designed to keep this detection time as short as possible. As shown in Figure 3, the first stage of the classifier employs a process of signal feature extraction, in which data are processed in order to extrapolate a number of signal features in the time and amplitude domains.

The second stage of the classifier is composed of four SVMs, which perform a first decision stage, outputting single binary partial hard decisions with respect to the related interference scenarios. The different SVMs are hereby described:

1. SVM-free channel: this SVM is trained to detect the presence of an IFC.
2. SVM-active network: targets the presence of an active IEEE 802.11 network occupying the related IEEE 802.15.4 PHY channel (i.e., an IEEE 802.11 access point with at least one associated terminal, generating uplink/downlink traffic).
3. SVM-silent network: targets a silent IEEE 802.11 network overlapping the specific channel. This is the case of an IEEE 802.11 access point with no associated terminal or an access point with associated terminals that are not generating data traffic in the observation time window.
4. SVM-microwave oven: detects the presence of RF leakage from a microwave-oven operating in close proximity to the radio node.

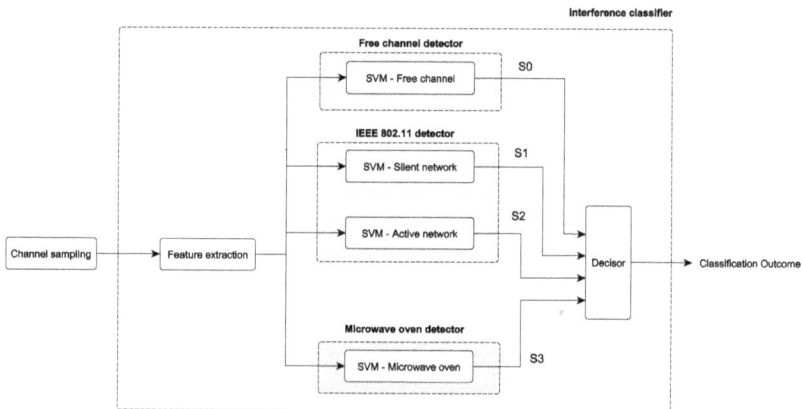

Figure 3. Setup of the support vector machine (SVM)-based interference classifier.

The outputs of the four SVM are represented by binary signals, S_1, S_2, S_3 and S_4, which have the value 1 (0) if the related decision is positive (negative). The binary decisions preformed by the SVMs are then processed by the logic decisor shown in Figure 4. The logic function of the decisor has been synthesized considering the cross-detection resilience of the single SVM. The final decision is composed of the four different classes listed in Table 1.

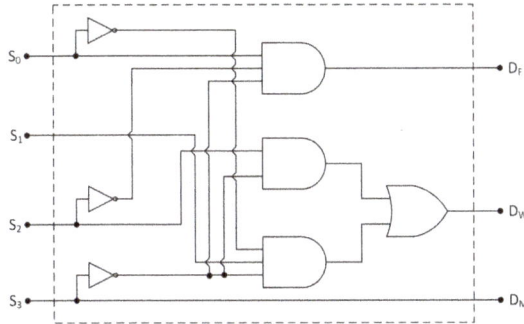

Figure 4. Details of the decisor for the proposed detection algorithm. The logic input signals are generated by the four SVMs in Figure 3.

Table 1. The four interference classes in the analysis.

D_F	D_W	D_M	Classification Outcome
1	0	0	The channel is free from the interference sources in the analysis.
0	0	1	A MWO was active during the sensing period.
0	1	0	An IEEE 802.11 network was overlapping the channel in the analysis.
0	0	0	The source of interference is unknown.

It must be highlighted that the classification performed based on the observation of a single radio node of the network only has local validity. This is because radio devices located in different locations of an industrial plant may be subjected to different interference conditions. In this context, the proposed method allows the interference scenario to be captured for each of the deployed nodes, opening the possibility of mapping the different sources of interference in the space-frequency domain. Nevertheless, since the aforementioned classification scheme exhibits a computational complexity that is beyond the capabilities of COTS WSN nodes, the proposed implementation relies on a centralized classification in place of a distributed approach. This, in turn, means that while the classifier can be implemented in the IWSAN network manager, the spectrum sensing and feature extraction process can be carried out by IWSAN nodes. This approach appears rather convenient, since, as described in Section 5.2, the signal processing required for the extraction of the selected signal features is kept to a minimum, while the efficiency of the classifier allows the radio nodes to work with small RSSI sample traces.

5.2. Signal Features

We select four signal features, belonging to two main classes: time domain and amplitude domain. The logic behind the selection is related to the properties of the interfering signals, such as the transmission airtime of IEEE 802.11 transmission and the periodicity of time domain pattern of the MWO RF leakage, as discussed in Section 3. To simplify the feasibility of the whole spectrum sampling and feature extraction processes on COTS WSN devices, our approach is to minimize the size of the RSSI trace, as well as the complexity of feature calculation methods.

5.2.1. Number and Length of Signal Bursts

The first time domain feature includes information about the burstiness of the observed signal, employing a threshold-based burst detection. The feature is an M-element vector in which each element represents the number of bursts of a certain sample length. Hence, we define:

$$\mathbf{F}_B = \{F_1,\ F_2,\ F_3,\ F_4,\ ..,\ F_M\} \tag{5}$$

where $F_n \in \mathbf{N}$ represents the the number of bursts of length n found in the RSSI trace in the analysis, while with F_M, we mark all of the bursts with sample length $L \geq M$. In particular, we require a certain number of samples under the selected threshold to identify the end of a burst. This is to avoid the case where a single or a few incorrect readings of the RSSI register will lead to a misclassification of long signal bursts into shorter ones. As will be discussed in Section 7, the detection of longer (i.e., > 5 ms) bursts is extremely important because it is a specific feature of the RF emissions of microwave ovens. The choice of a proper value of the threshold with respect to the calibration of the radio nodes will be discussed in Section 6.

5.2.2. Mean, Variance and Cardinality of Over-Threshold Samples

The second feature belongs to the amplitude domain and is defined as the mean value of the RSSI samples over the selected threshold θ. We define the vector containing all of the RSSI samples collected during the continuous observation window as $\mathbf{S} = \{s_0,\ s_1,\ s_2,\ ...,\ s_{N_S}\}$. Then, indicating with $\mathbf{S}^{(\mathbf{OT})}$ the subset of $\mathbf{S}$, such that $\mathbf{S}^{(\mathbf{OT})} = \{s_n \in \mathbf{S} \mid s_n > \theta\}$:

$$F_M = \frac{1}{N_{OT}} \sum_{i=1}^{N_{OT}} s_i^{(OT)} \tag{6}$$

with N_{OT} representing the cardinality of $\mathbf{S}^{(\mathbf{OT})}$, hence the number of above-threshold samples in the set. The third feature F_V follows directly from the definition of sample variance, hence using the same notation employed for F_M:

$$F_V = \frac{1}{N_{OT}} \sum_{i=1}^{N_{OT}} (s_i^{(OT)} - F_M)^2 \tag{7}$$

The last feature F_C counts the occurrences of RSSI samples above the threshold and hence is simply the cardinality of the set $\mathbf{S}^{(\mathbf{OT})}$. This feature nicely complements the previous two, adding information about the activity level of the interference source. It must be noted that while the signal features F_M, F_V and F_C are scalars, the feature F_B is an M-element vector; hence the SVM feature-space will be $M + 3$-dimensional, even using only four features.

6. Experimental Setup

6.1. Hardware Setup

The WSN devices selected for the experiments are Crossbow's TelosB motes CA2400 [39], equipped with Texas Instrument CC2420 transceiver [40]. The devices are programmed to collect a continuous set of RSSI samples with a sampling frequency of 2 kHz, over a sampling window that is selected according to the specific experiment. The RSSI value for each sample is fetched from the first 8 bits of register 0x16 of the CC2420 transceiver and represents the incident RF power in the selected 5 MHz-wide channel averaged over 128 μs (hence, eight IEEE 802.15.4 O-QPSK symbols). The RSSI data, fetched from the register in the form of an 8-bit signed integer, are buffered in the RAM and periodically saved to the internal flash memory. At the end of the sampling process, the content of the flash memory is sent over the USB port to a laptop, which logs the received data. The choice of this method for collecting RSSI data, in place of the direct sample-and-send over USB approach, was due to the insufficient bitrate (i.e., 115,200 baud including serial message overhead) available at the serial interface. In order to validate the performed measurements, we time stamp all of the observations and measure the delay of the instructions and task implemented in Tiny OS. This aspect will be further discussed in Section 8.2.2. Since Chen et al. [41] reported a consistent offset among RSSI readings performed with CC2420-equipped radio devices, we have also profiled our devices by means of a simple calibration process. We will discuss the effect of node calibration as well as the effect of this process on the performance of the proposed solution in Section 8.2.1.

6.2. Test Environments

The collection of experimental data has been carried out in three locations. Location A is a three-storey production plant employed for mineral processing. The environment is an open space cluttered with metal tanks, production machinery and a radio-controlled crane, while the three storeys are separated by metal grate flooring. A resident IEEE 802.11 WLAN covering the whole production plant was running at the time of the experiments.

Location B is a small mechanical workshop with an abundance of metal cluttering and soldering tools. A total of fourteen IEEE 802.11 access points with overlapping spectrum allocation on Channels {1, 6, 7, 11} is detectable in this environment. In Figure 5, we show the position of three of the nearest access points, which we label with AP1, AP2 and AP3, while the remaining devices were placed outside the range of the map, or on the upper floors of the building.

Location C is an office area, with nine IEEE 802.11 access points and residential MWO. We use this location to perform experiments on the classification of microwave interference, since neither industrial, nor residential MWO were present in the other two selected sites.

(a)   (b)

Figure 5. Two of the selected experimental environments. (**a**) Location A: industrial plant; (**b**) Location B: mechanical workshop.

6.3. The Collection of Training Data for SVM

As described in Section 4, the availability of a representative training dataset is fundamental for supervised-learning classification algorithms such as SVMs. Hence, particular attention has been put into building the dataset from both controlled and uncontrolled sources of interference and covering all 16 IEEE 802.15.4 channels.

6.3.1. Training Data from Uncontrolled IEEE 802.11 Networks

A preliminary set of measurements is collected in Location A, by means of Metageek channel analyzer [42], in order to determine the ground truth on the spectrum allocation of the resident IEEE 802.11 access points present at the industrial site. The network was composed of three IEEE 802.11 b/g/n access points statically allocated on IEEE 802.11 Channels 1, 6, and 11.

The training set is collected by means of a TelosB mote deployed in a fixed location of the industrial plant, programmed to sense each IEEE 802.15.4 channel for 10 min, collecting traces with over 1 M-sample per channel. Subsequently, the traces collected from the sampling of IEEE 802.11 Channels {15, 20, 25, 26} were assigned to the IFC class, since the IEEE 802.11 network did not overlap these channels (as shown in Figure 1), while the remaining traces were assigned to class IEEE 802.11 interference.

6.3.2. Training Data from Controlled Sources

The dynamics of an IEEE 802.11 network can vary greatly according to several factors (e.g., the number of connected devices and the traffic data-rate), and this in turn reflects the characteristics of the observed RSSI sample trace. While different methods for generating controlled interference are available in the literature, such as the one presented in [43], we use IEEE 802.11 hardware and a server-client architecture in order to have full control over the traffic distribution and transmission parameters. Following this approach, a controlled IEEE 802.11 network has been deployed at Location A. The structure of the network is represented in Figure 6 and is composed of a Linksys WRT610N IEEE 802.11 access point connected by an Ethernet cable to a Linux laptop running a traffic generator application generating the user datagram protocol (UDP) traffic with uniform, exponential and Pareto distributions. A second laptop, employing a Wi-Fi interface and running a Linux client, was used to receive and monitor the IEEE 802.11 packets. The access point was set on Channel 3, in order to overlap IEEE 802.15.4 Channel 15 (which was not affected by the resident industrial network), in order to isolate the observation from the effects of the resident network and capture only the effects of the custom IEEE 802.11 network.

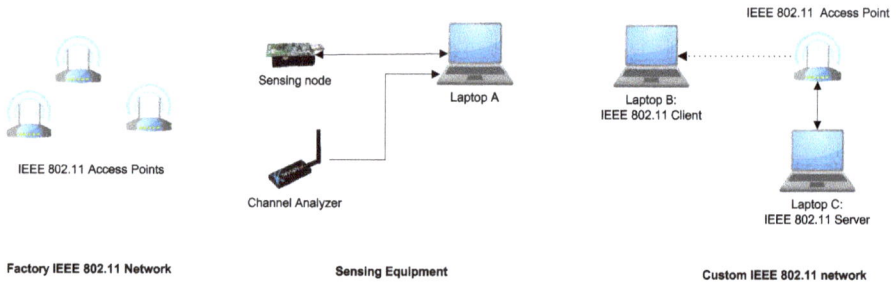

Figure 6. Experimental setup for the collection of training data for IEEE 802.11 interference detection.

6.3.3. Training Data from Microwave Oven

A set of measurements was collected in proximity (1 m) of a consumer Samsung MW82Y MWO set at the maximum heating power, achieving an active-passive heating phase with a duty cycle of 50%. The training data were collected along all of the IEEE 802.15.4 channels, since the temporal features of RF emissions from MWO can vary considerably moving within the 2.4-GHz ISM band due to the employed technology, as discussed in Section 3.

6.3.4. Test Data

We collected an extensive test dataset in the three described locations in order to thoroughly test the proposed algorithm.

At Location A, multiple RSSI traces from both the resident IEEE 802.11 network and the access point employed in the experiments were collected over all of the IEEE 802.15.4 channels. The traces were collected at several points of the three floors of the factory, taking care of including both line-of-sight (LOS) and non-line-of-sight NLOS propagation scenarios between the access points and the sensing node.

At Location B, we instead deployed the sensing node at one fixed point of the workshop (Point CS in Figure 5) and sensed each of the 16 channels for 5 min. At Location C, we deployed the radio node in the proximity of the active MWO, taking care of collecting measurements for all of the channels, randomizing the node position in the range 0.5 m–2 m from the oven. In all of the data collection points of the selected locations, the spectrum analyzer was used to determine the actual interference status of the sensed channels (similarly to the training data collection process), in order to determine the ground-truth for assessing the performance of the off-line classifier.

7. Results

7.1. Global Classification Accuracy

We tested the performance of the proposed algorithm by splitting each of the RSSI traces into several data chunks, with a length varying according to the tested sampling window, in the range of 50–500 ms. The data chunks were then processed in MATLAB, where we implemented the proposed classifier, including the feature extraction process and the four SVMs using the standard MATLAB SVM implementation with the Gaussian kernel, as well as the final decisor stage. For each test set, we calculated a detection accuracy metric by analyzing the outcome of the predicted interference source (according to Table 1) and comparing to the actual source determined during experiments. The detection accuracy was then simply calculated by dividing the number of correct classifications by the total number of classification rounds. In Table 2, we show the classification rates calculated for Locations A, B and C, including test data for all 16 IEEE 802.15.4 channels when the sampling widow is 250 ms; hence, chunks of 500 eight-bit RSSI samples are analyzed in each round of the test. It should be noted that the validity of the presented results is expected to be quite broad in nature as our dataset includes extensive traces from a broad range of scenarios including both controlled and uncontrolled IEEE 802.11 interference, spanning all of the IEEE 802.15.4 channels. A more in-depth discussion of the effects of a shorter or longer sampling window on the accuracy of the solution will be carried out in Section 8.1. In the following tables, we also include data about the distribution of misclassification in order to highlight which sources of interference were most likely to be misinterpreted by the classifier.

Table 2. Average classification accuracy for the 250 ms sampling window calculated over all of the scenarios. IFC, interference-free channel.

Channel Status	Detected Interference Source			
	IFC	IEEE 802.11	Microwave Oven	Unknown
IFC	91.2%	6.6%	2.1%	0.1%
IEEE 802.11	12.4%	83.9%	1.4%	2.3%
Microwave Oven	0.8%	16.3%	82.8%	0.1%

As shown in Table 2, the classifier was able to determine the presence of a free IEEE 802.15.4 channel 91.2% of the time, and the primary source of misclassification was the IEEE 802.11 network, which was detected 6.6% of the time. This fact is mainly due to the similarity of RSSI traces originating from an IEEE 802.11 network with low data traffic or even no associated terminal with RSSI originated from background noise. The similarity becomes more prominent when the signals originating from an IEEE 802.11 network and received by the WSN node are weak, due to attenuation effects. The same effect can also be used to explain the IFC misclassification rate of 12.4% when the interference comes from an IEEE 802.11 network. Nevertheless, in both cases, the introduction of the second support vector machine targeting silent IEEE 802.11 networks together with the employed decision logic helped to ensure a full-spectrum average detection accuracy of 83.9% for IEEE 802.11, even with a 250 ms sampling window, which is significantly shorter than other approaches presented in the literature (e.g., [21,26,28]). The classifier shows a detection accuracy of 82.8% when the source of interference was an MWO, where the most likely misclassification output was IEEE 802.11 interference, due to the similarities between temporal features of IEEE 802.11 and RF leakage from MWO. In Section 7.2, we point out that the detection accuracy appears to be significantly higher than the average for a specific contiguous set of IEEE 802.15.4 channels. This gives insight about dynamic channel-sensing strategies for maximizing the classification rate for this class of interference.

In Tables 3 and 4, we give more details about the full-spectrum detection accuracy from the datasets collected at Location A and Location B.

Table 3. Average classification accuracy for the 250 ms sampling window for Location A.

Channel Status	Detected Interference Source			
	IFC	IEEE 802.11	Microwave Oven	Unknown
IFC	98.2%	1.7%	0.1%	0.0%
IEEE 802.11	0.1%	98.9%	0.3%	0.7%

Table 4. Average classification accuracy for the 250 ms sampling window for Location B.

Channel Status	Detected Interference Source			
	IFC	IEEE 802.11	Microwave Oven	Unknown
IFC	84.9%	11.2%	3.8%	0.1%
IEEE 802.11	10.7%	77.9%	5.2%	6.1%

The classifier showed notable performances at Location A, being able to determine the correct source of interference 98.2% of the time when the channel was free and 98.9% when the RF emissions from the IEEE 802.11 network were overlapping the sensed channel. The average detection accuracy appeared lower at Location B, down to 77.9% for IEEE 802.11 interference. This is because the channel allocation of the resident IEEE 802.11 networks present at Location B was more challenging, including multiple overlapping networks with weaker signals and thus complicating the task of correctly classifying the interference on some WSN channels. This fact will be further analyzed in Section 7.2, where we provide in-channel detection accuracy analysis.

7.2. Channel-Specific Accuracy

In this section, the in-channel classification accuracy will be analyzed for all of the locations included in the tests. In Figure 7, we show the detection accuracy for Location A for a sampling window of 150 ms, together with information collected by means of the spectrum analyzer, showing the energy density in the 2.4-GHz ISM band at the industrial site. We chose to show the results for a shorter sampling window (150 ms) with respect to the previous section in order to highlight the impact of this aspect on the classification accuracy. As can be seen, even with this short sampling window, the detection rate ranged around 90% for the channels overlapped by the {1, 6, 11} configuration of the IEEE 802.11 network, while the IEEE 802.15.4 Channels {15, 20, 25, 26} were accurately reported free from interference.

In Figure 8, we show the classification outcome for Location B. In this test scenario, there are multiple IEEE 802.11 networks occupying IEEE 802.11 Channels 1, 6 and 11, while a distant access point with an average RSSI level < -80 dBm at the data collection point was present on Channel 7. As expected, while IEEE 802.15.4 Channels {15, 25, 26} were reported free in more then 85% of the tests, the decision for Channels 19 and 20 was uncertain since the IEEE 802.11 network was reported in only around 40–50% of the tests. We select this scenario to stress the performance of the proposed solution in the presence of an access point, which is barely detectable at the channel sensing location.

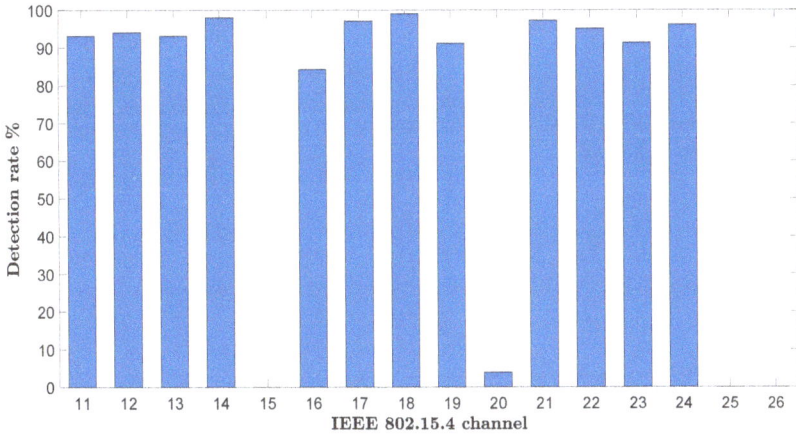

(a)

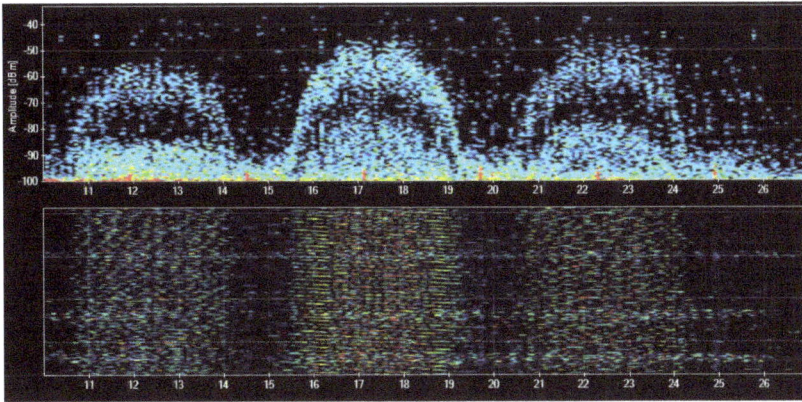

(b)

Figure 7. Location A, three IEEE 802.11 access point operating on IEEE 802.11 Channels {1, 6, 11}. (a) Channel-specific detection rate at Location A for the sampling window length of 150 ms; (b) IEEE 802.11 power spectral density (PSD) as observed by the channel analyzer.

In Figure 9, we show both the classification rate at Location C for a MWO and the average RSSI value and the standard deviation for the collected test data. As mentioned beforehand, while the RF leakage from MWO spans all along the 2.4-GHz ISM band, the detection accuracy presents considerable variations along the 16 IEEE 802.15.4 channels. In particular, the eight channels 16–23 seem to offer the best chance for microwave detection, while Channel 21 shows the maximum classification accuracy. This is because, as reported in [36] and the references therein, the residential MWOs have an emission peak frequency around 2.45 GHz, which corresponds to Channel 20 in the IEEE 802.15.4 mapping. In this case, we are likely to be experiencing an MWO with center emission frequency at 2.455 GHz, which consequently triggers a very high detection rate on Channel 21. Nevertheless, since the emission pattern may vary from model to model, the channel-specific performance is expected to vary from the one shown in Figure 9. In any case, for any model, the average detection accuracy is expected to remain consistent, since there will always be a region of maximum emission inside the ISM band of interest.

For the aforementioned reasons, a reasonable approach for channel sensing could be to perform the sensing on Channel 20 or on the adjacent channels in order to maximize the classification accuracy.

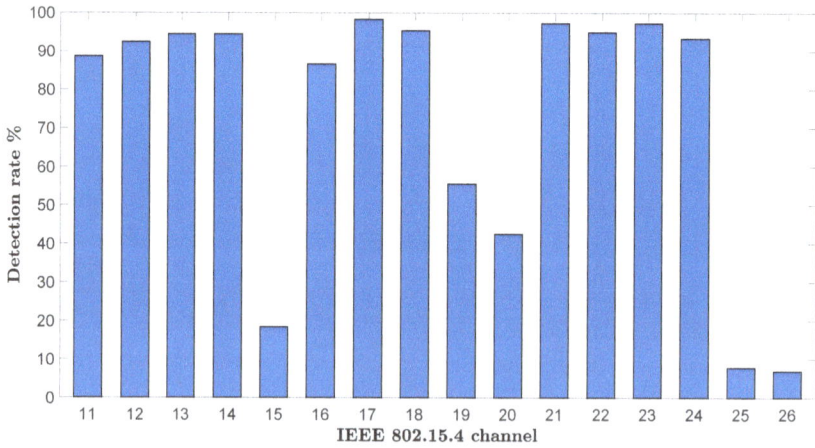

(a)

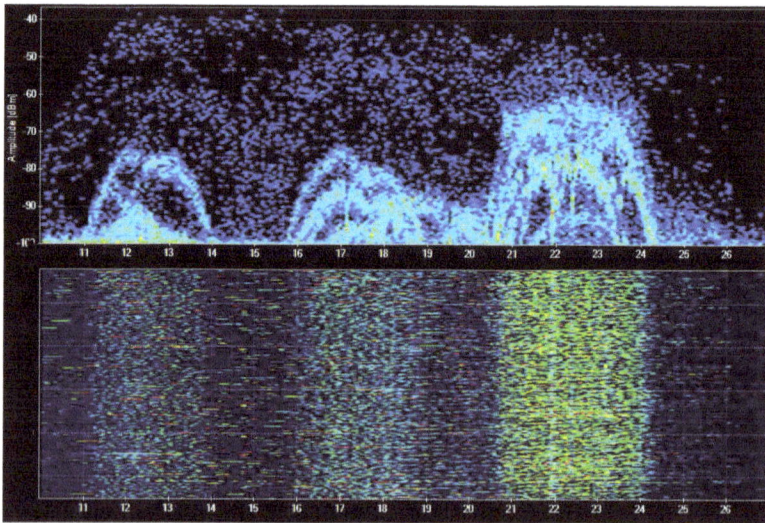

(b)

Figure 8. Location B, multiple IEEE 802.11 access points operating on Channels {1, 6, 7, 11}. (a) Channel-specific detection rate at Location B for a sampling window length of 150 ms.; (b) IEEE 802.11 power spectral density (PSD) as observed by the channel analyzer.

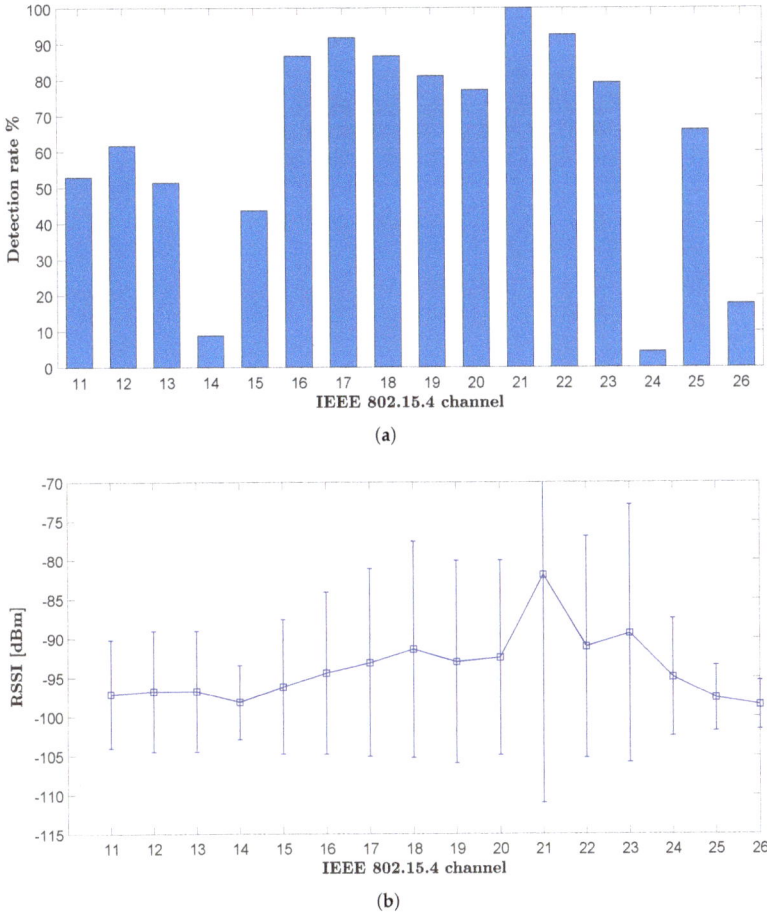

Figure 9. Location C, classification results for a sampling window length of 300 ms. (**a**) Channel-specific microwave oven detection rate; (**b**) average and standard deviation of RSSI traces collected in proximity of the tested microwave oven.

8. Discussion

8.1. The Influence of Sampling Window Length

In this section, we analyze the impact of different sampling window lengths on the classification accuracy of the three channel status classes of interest. The different sampling windows are tested by varying the number of samples included in the feature extraction process, as described in Section 5. In Figure 10, we show the curves for the average full-spectrum classification accuracy of IEEE 802.11 interference for sampling windows spanning from 50 ms–400 ms. We additionally show the curves representing the misclassification rate in order to highlight how the separation between classes is influenced by the sampling window.

It is interesting to note that the proposed classifier was not able to ensure proper separation between the classes IEEE 802.11 and IFC when the sampling window is $T_{SW} = 50$ ms, while the accuracy increases rapidly as T_{SW} approaches 200 ms, stabilizing around 84%. This behavior is mainly driven by the dynamics of IEEE 802.11 silent networks. Since a silent network shows by definition a low or null rate of exchanged data packets, due to a limited number of associated terminals, the on-air transmission is mainly due to the beacons emitted by the access point. Since the beacon period for all of the networks in experiments was set to the default value of 102.4 ms, a short sampling window can result in an increased possibility of missing the sensing of the beacon, which in turn reflects an insufficient separation between the vectors representing the IFC class and the IEEE 802.11 class in the employed $M + 3$-dimensional feature space. Nevertheless, thanks to the supervised-learning structure of the classifier, the proposed method allows the detection an IEEE 802.11 network with good accuracy in less than two beacon periods, while in concurrent approaches (e.g., [26]), the channel should be sensed for the time of several beacon periods in order to maximize the detection rate. The curves representing the classification rate for the IFCs are shown in Figure 11.

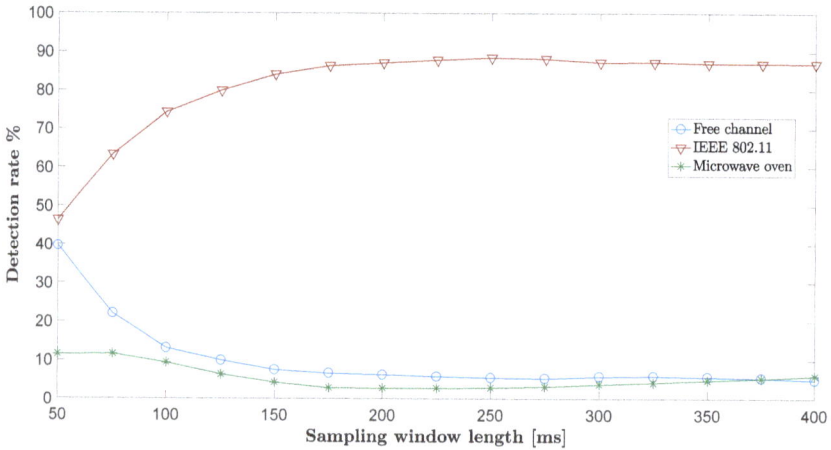

Figure 10. Average full-spectrum detection rate for IEEE 802.11 scenarios for different sampling window lengths.

In the case of IFC, the classification accuracy trend is opposite with respect to the IEEE 802.11 interference classification. This can be simply explained by the fact that shorter observation windows will in turn mean a lower probability of encountering amplitude fluctuations of the background noise, which can potentially drift the feature vector in the decisional zone of IEEE 802.11 and MWO classes. Despite this fact, the IFC classification rate was reported consistently above 90%, even for sampling windows greater than 200 ms, while we observed an increase of MWO and IEEE 802.11 misclassification, for the reason just described.

In Figure 12, we show the full-spectrum classification accuracy for MWO interference.

The figure shows insufficient separation between the classes MWO and IEEE 802.11 for $T_{SW} = 50$ ms, while increasing the sampling time improves the classification accuracy, even if the improvement is significantly slower with respect to the case of IEEE 802.11 interference. This in turn means that in order to maximize the separation between classes, a sampling window of $T_{SW} \geq 250$ ms is required so that the selected features can emerge with sufficient clarity and ensure a full spectrum classification accuracy greater than 82%. As discussed in Section 7, this behavior is due to the similarity of the temporal features of IEEE 802.11 signals in the case of active networks and the RF leakage of MWO.

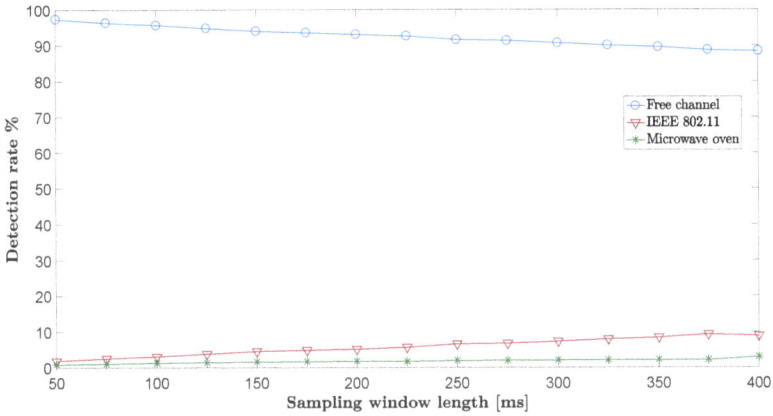

Figure 11. Average full-spectrum detection rate for interference-free scenarios for different sampling window lengths.

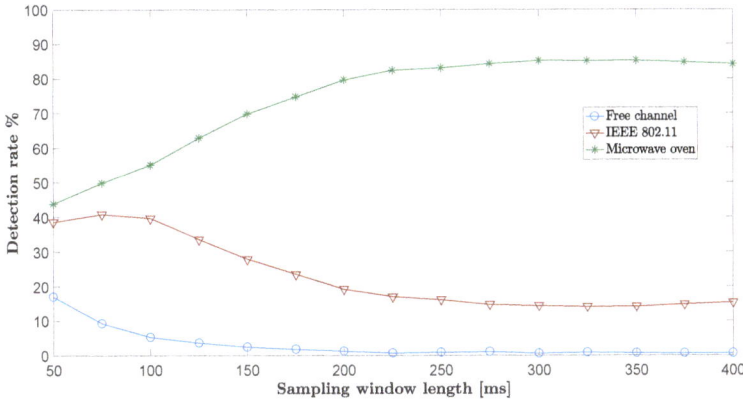

Figure 12. Average full-spectrum detection rate for interference from MWO for different sampling window lengths.

In Figure 13, we finally show the effects of different sampling windows on the in-channel detection accuracy for MWO at Location C.

From the plot, we observe that the classification accuracy is monotonically increasing for all of the channels, meaning that longer sampling windows are always beneficial for MWO detection. It can also be noted that the classification accuracy dip on Channel 14 and on Channel 24 experiences a significant improvement when the sampling window approaches 350 ms, giving a hint about the bursty time distribution of the RSSI samples on the side spectrum of MWO leakage.

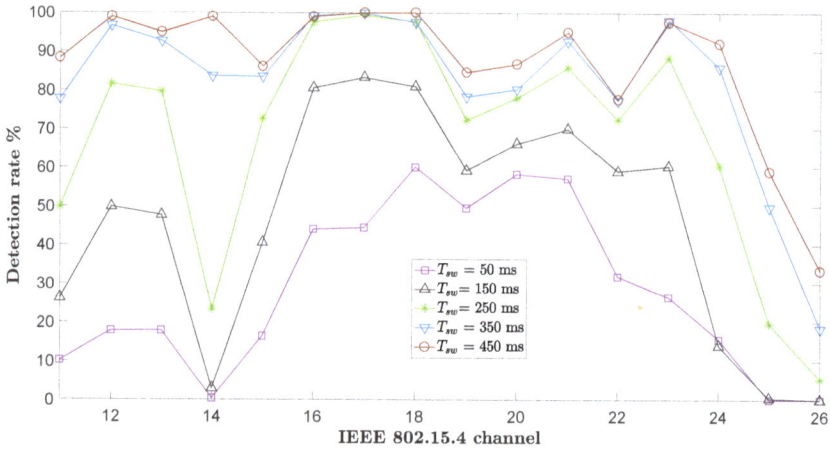

Figure 13. Detection rate for microwave oven at Location C for different sampling window lengths. Some curves have been removed for clarity.

8.2. Hardware-Related Considerations

Since COTS WSN nodes are low-power devices with resource-constrained hardware, particular attention has to be paid when implementing a complex methods on these platforms. In this section, we discuss how certain characteristics of the selected hardware (i.e., CC2420-equipped TelosB motes) influence the spectrum-sensing task and consequently the applicability and performance of the proposed classification method.

8.2.1. The Role of Node Calibration

It is a well-known fact that different CC2420-based devices can show variation in the nominal response of the RSSI curve. Since the core of the proposed method is based on RSSI sampling and threshold-based features, it is of primary importance to analyze if these variations can hamper the performance of the classifier. In their work, Chen et al. [41] showed that these variations are due to two different phenomena: a non-linearity in the CC2420 RSSI response curve and the presence of a node-dependent offset. While the first phenomenon is of minor relevance, since the non-linear and non-injective regions do not influence significantly the RSSI curve (which remains mostly linear), a consistent offset of ±6 dB is reported among different nodes.

We have tested several different TelosB nodes, sampling IEEE 802.15.4 Channel 26 in a radio-controlled environment to determine both the amplitude distribution and the mean of the collected RSSI traces for different nodes. As shown in Figure 14, we have observed a maximum RSSI offset of ±5 dB.

We carried out an analysis of the influence of the RSSI offset existing between the network device deployed for channel sensing and the device used for preliminary training set collection. In Figure 15, we show the impact of RSSI offset on the classification accuracy of the three targeted interference classes.

As can be observed, performing the channel sensing using a node with a consistent RSSI offset can greatly hamper the performance of the classifier, also considering that the offset between two nodes could theoretically span up to 12 dB. Even an offset of 5 dB, such as the one reported in our node set, can decrease the performance of both IEEE 802.11 andMWO classification up 15–20%, rendering a node calibration process a relevant step from the perspective of safeguarding the performance of the proposed approach. Fortunately, this process is straightforward, since as shown in [41], only a simple noise-floor-based RSSI offset calculation and compensation is needed. In the proposed approach, for example, once the RSSI offset is acquired, the offset compensation can be simply implemented by employing a software-based adaptation of the energy-threshold used for the feature-extraction task.

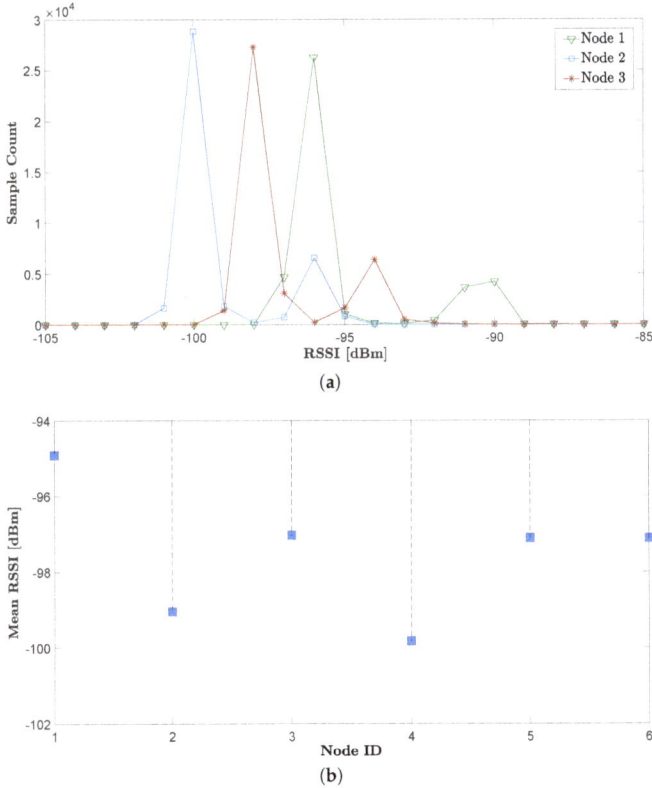

(a)

(b)

Figure 14. The RSSI profiling process for TelosB motes. (**a**) Amplitude distribution of RSSI traces from background noise sensing for different CC2420 nodes. Some curves have been removed for clarity. (**b**) Mean of recorded RSSI sample traces.

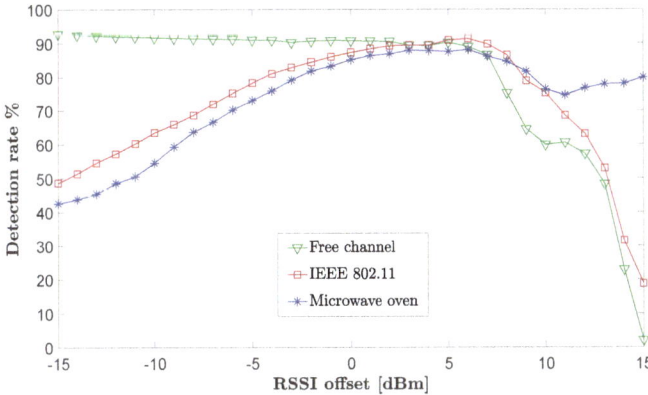

Figure 15. Detection accuracy with respect to RSSI offset between the node used for collecting training data and the actual sensing node.

8.2.2. Assessing the Timeliness of the Sampling Process

In order to discuss the feasibility of the proposed sensing scheme with respect to the employed COTS hardware platform, we monitor and analyze the delay generated by the various operations needed to perform the in-node channel sensing. In Figure 16, we show the partial duration of the tasks implemented in TelosB motes in order to acquire and store the RSSI samples. Two of the most demanding tasks in terms of delay are the request for accessing and releasing the I/O resources, requiring 212 µs and 74 µs, respectively. In addition, the tasks of setting the CSn (chip select) pin for reading the CC2420 RSSI register lasts 12 µs, while the actual operation of sampling the value of RSSI register takes 112 µs to be completed.

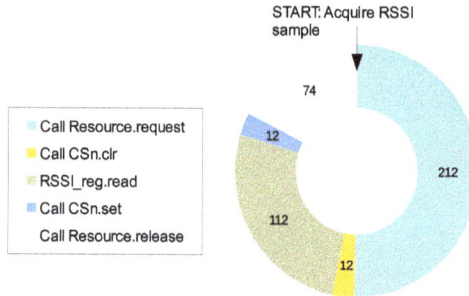

Figure 16. Operation delays in µs for the channel-sensing task implemented in TelosB motes.

The total delay for collecting and storing one sample is then 429 µs, while we use a sampling frequency of 2 kHz, corresponding to a 500 µs sampling period. With the current approach, the sampling frequency could be theoretically pushed up to 4.6 kHz if the CC2420 resources are not released until the end of the whole sampling process. In other approaches (e.g., [26]), the implementation for the channel sensing in Contiki OS allows for a sampling rate up to 8.13 kHz. Nevertheless, a higher sampling frequency means more data to process, as well as a more stressful and energy-consuming sampling process. Therefore, in this work, we have employed a more relaxed sampling timing, while we rely on the approach of an advanced classification algorithm, in order to maintain high classification performance while ensuring a lower memory footprint.

9. Conclusions

In this paper, we present a novel scheme employing machine learning methods for cross-technology interference classification in IWSAN. The proposed method employs a three-step classifier composed of a lightweight feature-extraction process, a preliminary classification stage employing four SVMs and a final decisor, allowing for classification among interference from IEEE 802.11 networks, microwave ovens, as well as the presence of interference-free channels. The tests conducted in industrial environments, including a wide range of interference scenarios, show an average classification accuracy of 84% and up to 98% for IEEE 802.11 active networks, with a channel sensing time of 300 ms. The memory footprint of the samples collected in this sensing time remains below 600 bytes per channel thanks to the limited sampling frequency. The extremely short time required for sensing renders the developed solution a promising candidate for the adoption in superframe-based TSCH networks by means of spectrum-sensing-reserved timeslots. In this paper, we have also highlighted the fundamental influence of device calibration on the performance of spectrum-sensing-based methods using COTS WSN hardware, which is a matter often overlooked in related literature. In particular, it is shown that the classification accuracy of the proposed solution is significantly influenced by the intrinsic hardware variations.

We leave to future works further investigations on the potentialities of SVM-based-methods for interference classification in IWSAN. Other notable aspects of interest are the inclusion of the channel-sensing and classification mechanism in a TSCH network and a run-time assessment of the solution, as well as the development of interference mitigation strategies.

Acknowledgments: The authors would like to thank Imerys Mineral AB for the access to their production plant in Sundsvall and R.Rondón from Mid Sweden University for the helpful feedback.

Author Contributions: S.G. conceived and implemented the proposed method; S.G and A.M. designed and conducted the experiments; S.G. analyzed the data. M.G. supervised the overall work. S.G., A.M. wrote the paper. All authors contributed in discussing and revising the manuscript.

Conflicts of Interest: The authors declare no conflict of interest.

References

1. Prathap, U.; Shenoy, P.D.; Venugopal, K.R.; Patnaik, L.M. Wireless Sensor Networks Applications and Routing Protocols: Survey and Research Challenges. In Proceedings of the 2012 International Symposium on Cloud and Services Computing, Mangalore, India, 17–18 December 2012; pp. 49–56.
2. Đurišić, M.P.; Tafa, Z.; Dimić, G.; Milutinović, V. A survey of military applications of wireless sensor networks. In Proceedings of the 2012 Mediterranean Conference on Embedded Computing (MECO), Bar, Montenegro, 19–21 June 2012; pp. 196–199.
3. Mangali, N.K.; Kota, V.K. Health monitoring systems: An energy efficient data collection technique in wireless sensor networks. In Proceedings of the 2015 International Conference on Microwave, Optical and Communication Engineering (ICMOCE), Bhubaneswar, India, 18–20 December 2015; pp. 130–133.
4. Benkhelifa, I.; Nouali-Taboudjemat, N.; Moussaoui, S. Disaster Management Projects Using Wireless Sensor Networks: An Overview. In Proceedings of the 2014 28th International Conference on Advanced Information Networking and Applications Workshops, Victoria, BC, Canada, 13–16 May 2014; pp. 605–610.
5. *IEEE Standard for Information technology—Local and metropolitan area networks—Specific requirements—Part 15.4: Wireless Medium Access Control (MAC) and Physical Layer (PHY) Specifications for Low Rate Wireless Personal Area Networks (WPANs)*; IEEE Std 802.15.4-2006, IEEE: New York, NY, USA; 2006.
6. Åkerberg, J.; Gidlund, M.; Björkman, M. Future research challenges in wireless sensor and actuator networks targeting industrial automation. In Proceedings of the 2011 9th IEEE International Conference on Industrial Informatics, Lisbon, Portugal, 26–29 July 2011; pp. 410–415.
7. Galloway, B.; Hancke, G. Introduction to Industrial Control Networks. *IEEE Commun. Surv. Tutor.* **2013**, *15*, 860–880.
8. Rappaport, T.S. Indoor radio communications for factories of the future. *IEEE Commun. Mag.* **1989**, *27*, 15–24.
9. Gwarek, W.K.; Celuch-Marcysiak, M. A review of microwave power applications in industry and research. In Proceedings of the 15th International Conference on Microwaves, Radar and Wireless Communications (IEEE Cat. No.04EX824), Warsaw, Poland, 17–19 May 2004; Volume 3, pp. 843–848.
10. Sikora, A. *Wireless Personal and Local Area Networks*; Wiley: Chichester, UK; Hoboken, NJ, USA, 2003.
11. Liang, C.J.M.; Priyantha, N.B.; Liu, J.; Terzis, A. Surviving Wi-Fi Interference in Low Power ZigBee Networks. In Proceedings of the 8th ACM Conference on Embedded Networked Sensor Systems, Zürich, Switzerland, 3–5 November 2010; ACM: New York, NY, USA, 2010; pp. 309–322.
12. Yang, D.; Xu, Y.; Gidlund, M. Wireless Coexistence between IEEE 802.11- and IEEE 802.15.4-Based Networks: A Survey. *Int. J. Distrib. Sens. Netw.* **2011**, *7*, 912152.
13. Hermans, F.; Rensfelt, O.; Voigt, T.; Ngai, E.; Norden, L.A.; Gunningberg, P. SoNIC: Classifying Interference in 802.15.4 Sensor Networks. In Proceedings of the 12th International Conference on Information Processing in Sensor Networks (IPSN '13), Philadelphia, PA, USA, 8–11 April 2013; ACM: New York, NY, USA, 2013; pp. 55–66.
14. *HART Communication Protocol Specification, Revision 7.4*; Technical Report; HART Communication Foundation: Austin, TX, USA, 2012.
15. *Wireless Systems for Industrial Automation: Process Control and Related Applications*; ISA 100.11a-2011; International Society of Automation: Research Triangle Park, NC, USA, 2011.

16. *Industrial Communication Networks—Fieldbus Specifications—WIA-PA Communication Network and Communication Profile*; IEC 62601; International Electrotechnical Commission: Geneva, Switzerland, 2011.

17. Yucek, T.; Arslan, H. A survey of spectrum sensing algorithms for cognitive radio applications. *IEEE Commun. Surv. Tutor.* **2009**, *11*, 116–130.

18. Brown, J.; Roedig, U.; Boano, C.A.; Römer, K. Estimating packet reception rate in noisy environments. In Proceedings of the 39th Annual IEEE Conference on Local Computer Networks Workshops, Edmonton, AB, Canada, 8–11 September 2014; pp. 583–591.

19. Zacharias, S.; Newe, T.; O'Keeffe, S.; Lewis, E. A Lightweight Classification Algorithm for External Sources of Interference in IEEE 802.15.4-Based Wireless Sensor Networks Operating at the 2.4 GHz. *Int. J. Distrib. Sens. Netw.* **2014**, *10*, 265286.

20. Zhou, R.; Xiong, Y.; Xing, G.; Sun, L.; Ma, J. ZiFi: wireless LAN discovery via ZigBee interference signatures. In Proceedings of the sixteenth annual international conference on Mobile computing and networking, Chicago, IL, USA, 20–24 September 2010; ACM: New York, NY, USA, 2010; pp. 49–60.

21. Gao, Y.; Niu, J.; Zhou, R.; Xing, G. ZiFind: Exploiting cross-technology interference signatures for energy-efficient indoor localization. In Proceedings of the 2013 Proceedings IEEE INFOCOM, Turin, Italy, 14–19 April 2013; pp. 2940–2948.

22. Choi, J. WidthSense: Wi-Fi Discovery via Distance-based Correlation Analysis. *IEEE Commun. Lett.* **2016**, *21*, 422–425.

23. Petrova, M.; Wu, L.; Mahonen, P.; Riihijarvi, J. Interference Measurements on Performance Degradation between Colocated IEEE 802.11 g/n and IEEE 802.15.4 Networks. In Proceedings of the Sixth International Conference on Networking (2007. ICN '07), Sainte-Luce, France, 22–28 April 2007; p. 93.

24. Hossian, M.M.A.; Mahmood, A.; Jäntti, R. Channel ranking algorithms for cognitive coexistence of IEEE 802.15.4. In Proceedings of the 2009 IEEE 20th International Symposium on Personal, Indoor and Mobile Radio Communications, Tokyo, Japan, 13–16 September 2009; pp. 112–116.

25. ZigBee Standards Organization. *ZigBee Specifications*; ZigBee Standards Organization: San Ramon, CA, USA, 2012; pp. 1–622.

26. Zacharias, S.; Newe, T.; O'Keeffe, S.; Lewis, E. 2.4 GHz IEEE 802.15.4 channel interference classification algorithm running live on a sensor node. In Proceedings of the 2012 IEEE Sensors, Taipei, Taiwan, 28–31 Octorber 2012; pp. 1–4.

27. Iyer, V.; Hermans, F.; Voigt, T. Detecting and Avoiding Multiple Sources of Interference in the 2.4 GHz Spectrum. In Proceedings of the 12th European Conference on Wireless Sensor Networks (EWSN), Porto, Portugal, 9–11 February 2015; Abdelzaher, T., Pereira, N., Tovar, E., Eds.; Springer International Publishing: Cham, Switzerland, 2015; pp. 35–51.

28. Ansari, J.; Ang, T.; Mähönen, P. WiSpot: fast and reliable detection of Wi-Fi networks using IEEE 802.15.4 radios. In Proceedings of the 9th ACM International Symposium on Mobility Management and Wireless Access, Miami, FL, USA, 31 October–4 November 2011; ACM: New York, NY, USA, 2011; pp. 35–44.

29. Chowdhury, K.R.; Akyildiz, I.F. Interferer Classification, Channel Selection and Transmission Adaptation for Wireless Sensor Networks. In Proceedings of the 9th IEEE International Conference on Communications (ICC '09), Dresden, Germany, 14–18 June 2009; pp. 1–5.

30. Rayanchu, S.; Patro, A.; Banerjee, S. Airshark: Detecting non-WiFi RF Devices Using Commodity WiFi Hardware. In Proceedings of the 2011 ACM SIGCOMM Conference on Internet Measurement Conference (IMC '11), Berlin, Germany, 2–4 November 2011; ACM: New York, NY, USA, 2011; pp. 137–154.

31. Weng, Z.; Orlik, P.; Kim, K.J. Classification of wireless interference on 2.4 GHz spectrum. In Proceedings of the 2014 IEEE Wireless Communications and Networking Conference (WCNC), Istanbul, Turkey, 6–9 April 2014; pp. 786–791.

32. Hermans, F.; Larzon, L.A.; Rensfelt, O.; Gunningberg, P. A Lightweight Approach to Online Detection and Classification of Interference in 802.15.4-based Sensor Networks. *SIGBED Rev.* **2012**, *9*, 11–20.

33. Nicolas, C.; Marot, M. Dynamic link adaptation based on coexistence-fingerprint detection for WSN. In Proceedings of the 2012 11th Annual Mediterranean Ad Hoc Networking Workshop (Med-Hoc-Net), Ayia Napa, Cyprus, 19–22 June 2012; pp. 90–97.

34. Barać, F.; Gidlund, M.; Zhang, T. Ubiquitous, Yet Deceptive: Hardware-Based Channel Metrics on Interfered WSN Links. *IEEE Trans. Veh. Technol.* **2015**, *64*, 1766–1778.

35. Zheng, X.; Cao, Z.; Wang, J.; He, Y.; Liu, Y. ZiSense: Towards Interference Resilient Duty Cycling in Wireless Sensor Networks. In Proceedings of the 12th ACM Conference on Embedded Network Sensor Systems, (SenSys '14), Memphis, TN, USA, 3–6 November 2014; ACM: New York, NY, USA, 2014; pp. 119–133.

36. Rondeau, T.W.; D'Souza, M.F.; Sweeney, D.G. Residential microwave oven interference on Bluetooth data performance. *IEEE Trans. Consum. Electron.* **2004**, *50*, 856–863.

37. Bishop, C.M. *Pattern Recognition and Machine Learning (Information Science and Statistics)*; Springer: Secaucus, NJ, USA, 2006.

38. Boyd, S.; Vandenberghe, L. *Convex Optimization*; Cambridge University Press: New York, NY, USA, 2004.

39. Crossbow TelosB Mote Plattform, Datasheet. Available online: http://www.willow.co.uk/TelosB_Datasheet.pdf (accessed on 16 June 2017).

40. Texas Instruments CC2420 - 2.4 GHz IEEE 802.15.4/ZigBee-ready RF Transceiver. Available online: http://www.ti.com/lit/ds/symlink/cc2420.pdf (accessed on 16 June 2017).

41. Chen, Y.; Terzis, A. On the mechanisms and effects of calibrating RSSI measurements for 802.15.4 radios. In Proceedings of the 7th European conference on Wireless Sensor Networks, Coimbra, Portugal, 17–19 February 2010; Springer: Berlin/Heidelberg, Germany, 2010; pp. 256–271.

42. Metageek Wi-Spy Chanalizer. Available online: http://files.metageek.net/marketing/data-sheets/MetaGeek_Wi-Spy-Chanalyzer_DataSheet.pdf (accessed on 16 June 2017).

43. Boano, C.A.; Voigt, T.; Noda, C.; Römer, K.; Zúñiga, M. JamLab: Augmenting sensornet testbeds with realistic and controlled interference generation. In Proceedings of the 10th ACM/IEEE International Conference on Information Processing in Sensor Networks, Chicago, IL, USA, 12–14 April 2011; pp. 175–186.

Journal of
Sensor and Actuator Networks

MDPI

Article

Dynamic Cooperative MAC Protocol for Navigation Carrier Ad Hoc Networks: A DiffServ-Based Approach

Chao Gao [1,2,*], Bin Zeng [1], Jianhua Lu [1] and Guorong Zhao [1]

[1] Department of Control Engineering, Naval Aeronautical and Astronautical University, Yantai 264001, China; 15966531553@163.com (B.Z.); llljjjhua001@sina.com (J.L.); GRZhao6881@163.com (G.Z.)
[2] The 92664th Unit of PLA, Qingdao 266000, China
*   Correspondence: gaochao.shd@163.com; Tel.: +86-153-3545-7310

Received: 21 October 2016; Accepted: 26 July 2017; Published: 8 August 2017

Abstract: In this paper, a dynamic cooperative MAC protocol (DDC-MAC) based on cluster network topology is proposed, which has the capability of differentiated service mechanisms and long-range communication. In DDC-MAC, heterogeneous communications are classified according to service types and quality of service (QoS) requirements, i.e., periodic communication mode (PC mode) is extracted with a QoS guarantee for high-frequency periodic information exchange based on adapt-TDMA mechanisms, while other services are classified as being in on-demand communication mode (OC mode), which includes channel contention and access mechanisms based on a multiple priority algorithm. OC mode is embedded into the adapt-TDMA process adaptively, and the two communication modes can work in parallel. Furthermore, adaptive array hybrid antenna systems and cooperative communication with optimal relay are presented, to exploit the opportunity for long-range transmission, while an adaptive channel back-off sequence is deduced, to mitigate packet collision and network congestion. Moreover, we developed an analytical framework to quantify the performance of the DDC-MAC protocol and conducted extensive simulation. Simulation results show that the proposed DDC-MAC protocol enhances network performance in diverse scenarios, and significantly improves network throughput and reduces average delay compared with other MAC protocols.

Keywords: Navigation Carrier Ad hoc Network (NC-NET); MAC protocol; differentiated services (DiffServ); cooperative communication; optimal relay

---

1. Introduction

Emerging advantages of network technique inspire other fields to solve their own bottlenecks through networked approaches [1]. In our research, we are devoted to putting forward a network-aware solution for a bottleneck of modern navigation technology, which prevents it from reaching a higher level [1–3]. More specifically, the navigation precision of each navigation technology is closely related to the self-contained degree of the navigation system; on one hand, we can upgrade its integrity to enhance the navigation precision, e.g., by augmenting other navigation equipment or replacing it with a high precision navigation system, which will lead to additional economic investment and physical load; on the other hand, we can simplify the complexity of the navigation equipment to alleviate the burdens above, e.g., by the reduction of navigation equipment or its replacement with a low-cost navigation system, although its navigation capacity will be degraded accordingly. In references [2–4], we proposed a navigation-oriented network—navigation carrier ad hoc networks (NC-NET)—to solve the localization, target tracking, and multimedia data exchange problems through a network approach. The proposed NC-NET, which is essentially an ad hoc network between navigation carriers (NCs), is surveyed as a new network family. Moreover, some of the protocols and mechanisms have already been investigated, including (i) the protocol framework, and the models in the physical layer [2,3]; (ii) a distributed multi-weight data-gathering and aggregation protocol for cluster topology, i.e., DMDG protocol [2]; (iii) a grid-based cooperative QoS routing protocol with fading memory optimization, i.e., FMCQR protocol [3]; and (iv) a network-based localization mechanism [4]. As part of a series of research work, the main objective of this paper is to develop a MAC protocol with full account of the features of NC-NET.

Figure 1 illustrates a typical use of NC-NET, which is a scenario of multi-service transmission among a set of unmanned aerial vehicles (UAVs); herein, each UAV is equipped with an NC-NET terminal with a unique ID, and thus the objective node s will exchange the wireless packets with its neighbors $C_{M(i)}, i \in \{1, 2, \ldots, 10\}$ at each round, according to the uniform NC-NET protocols. Thus, in this kind of application, an efficient MAC protocol is indispensable for NC-NET, which has the following design challenges: (i) multiple traffic services; (ii) three-dimensional large-scale active region (e.g., aerial, ground or water surface), and long-range distance between nodes (e.g., line-of-sight, beyond-line-of-sight); and (iii) heterogeneous types of NC-NET nodes. For simplicity, but without loss of generality, an essential NC-NET is discussed at length in the research, in which the open issues surrounding general-purpose MAC protocols are analyzed in detail. However, there are also several common points between NC-NET and the previous ad hoc network, e.g., decentralization, infrastructure-less, and self-configuration [2,3,5]. Moreover, several illustrations of NC-NET adopt the general-purpose technologies of ad hoc networks, such as TDMA and CSMA/CA. From this point of view, NC-NET can also be investigated as a special form of ad hoc network. In previous literature, several effective technologies for MAC protocols have been studied in depth, which can be used as a reference for our research, e.g., directional antennas, multi-channel diversity, differentiated-service mechanisms, and cooperative communication.

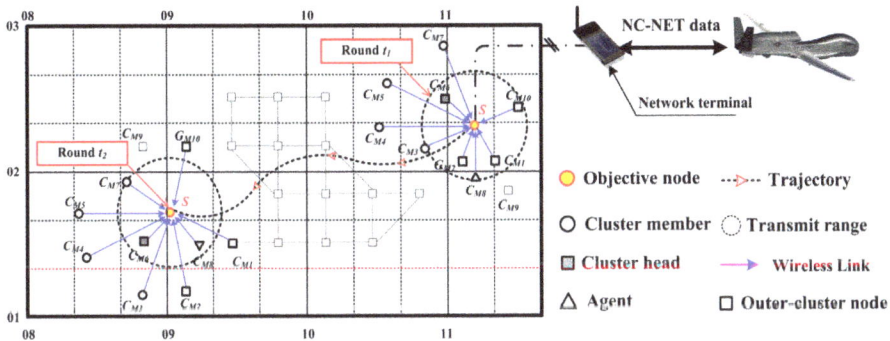

Figure 1. Typical illustration of networked UAVs based on NC-NET.

There are several common points between NC-NET and previous ad hoc networks, such as decentralization, lack of infrastructure, and self-configuration [6,7]. Moreover, several illustrations of NC-NET, e.g., tactical targeting network technology (TTNT) [8], have adopted the general-purpose technologies of ad hoc network such as TDMA and CSMA/CA. From this point of view, NC-NET can also be investigated as a special form of ad hoc network. In previous literature, several effective technologies for MAC protocols of ad hoc network have been studied in depth, and can be used as a reference for our research, i.e., directional antennas [9–11], multi-channel diversity [12], differentiated-service mechanisms [13–18], and cooperative communication [19–24].

Following detailed analysis, it is clear that the results for challenge (i), challenge (ii), and their hybrid are extensive; however, in terms of challenge (iii), the integrated investigations on challenges (i) and (ii) are quite limited. From the above analysis, we identify the missing properties in the existing work for QoS provisioning in NC-NET and introduce the design and implementation of a new MAC protocol for NC-NET. In this paper, we propose a hybrid approach, namely, a DiffServ-based dynamic cooperative MAC (DDC-MAC) protocol, in combination with the aforementioned techniques for NC-NET. The key contributions of this work and highlights of DDC-MAC are listed as follows: First, a system model is presented that exploits the merits of the aforementioned issues, including an adaptive array hybrid antenna system, multi-channel environment and cluster-based network topology. Second, the proposed DDC-MAC protocol is designed in detail with the following features: (i) a DiffServ mechanism is proposed to realize the parallel transmission of multiple services with QoS guarantee; (ii) a multi-priority policy is presented to reduce the probability of contention and collision, and to reassign the access sequence; (iii) an adaptive channel back-off sequence (ACBS) is proposed as a reference for lower-priority pairs to switch directly among data channels without additional negotiation packets; and (iv) an optimal-relay-based MAC procedure is deduced to handle link failures in long-range transmission. The performance of DDC-MAC is theoretically analyzed and evaluated. Finally, the results demonstrate the efficiency of DDC-MAC by comparing it with other contemporary protocols.

The remainder of the paper is organized as follows. Section 3 presents the system description and traffic models. Section 4 provides a detailed description of the proposed DDC-MAC protocol. In Section 5, we analyze DDC-MAC in terms of the searching probability of next-hop nodes and saturated throughput. In Section 6, we evaluate DDC-MAC based on extensive simulations. Finally, conclusions are given in Section 7.

2. Related Works

Although extensive recent papers delve specifically into QoS-aware MAC problems, the literature pertaining to QoS MAC for high-dynamic ad hoc networks is extremely limited, and that addressing NC-NET remains almost non-existent, in spite of being described as an open issue at the link layer in reference [1]. In this section, we analyze the related works on typical QoS-aware MAC protocols for wireless networks.

Directional antennas can offer clear advantages in terms of improving the signaling range without extra power (as opposed to omni-directional antennas), especially for long-range transmission. Previous research has incorporated directional antennas in the design of MAC protocols. DMAC [5] is one of the earliest of these. This protocol is based on a modified 802.11 MAC protocol [6], and uses a per-sector blocking mechanism to block a sector once it senses a request-to-send (RTS) or clear-to-send (CTS) packet, whereas the data packets are transmitted through directional beams. This kind of mechanism is beneficial for spatial reuse, end-to-end transmission delay, and power consumption in data transmission [7]. Unfortunately, some problems also arise from the use of directional antennas, such as the deafness problem [8], the new hidden terminal problem [8,9], and the head-of-line (HOL) blocking problem [10]. Previous directional MAC protocols have already defined several techniques to handle these problems [8–12]. However, they mainly focus on some specific aspects; for example, augmentation of the busy-tone channel [10,11], attachment of a start-of-frame [8,9], or multi-channel diversity [12,13]. Nevertheless, the question of how to combine these techniques in a large-scale mobile wireless network and maximize the advantages of directional antennas in a multi-service environment is still an open issue, motivating the research in this study. Furthermore, in view of the complex communication challenges in NC-NET, in this research, the usage of directional antennas is combined with other techniques, i.e., differentiated services and cooperative communication.

Differentiated services (DiffServ) is a widely known and utilized technique for QoS provision in networks with multiple traffic services [14], which are differentiated and prioritized based on one or more criteria and formed into several traffic classes. In the literature [14–18], the majority of DiffServ mechanisms are proposed for wireless multimedia sensor networks (WMSN), and can be classified according to various types of criteria, such as drive-mode-aware criteria, QoS-aware criteria, and other criteria. In references [15,16], the services are differentiated based on drive modes, whereby the cyclically occurring services are classified as a time-driven class, whereas stochastic services are classified as an event-driven class. Thereby, the medium access of the two classes is realized based on schedule-based MAC and contention-based MAC, respectively, or hybrid solutions including TDMA/CSMA, CDMA/CSMA, FDMA/CSMA, etc. In references [17,18], multiple services are classified based on QoS-aware criteria such as delay sensitivity and packet loss sensitivity. Saxena et al. [17] proposed a CSMA/CA approach, classifying the co-existing packets into two categories, i.e., streaming multimedia (over UDP) packets and Best Effort FTP (over TCP) packets. Moreover, Diff-MAC [18] is designed with the key features of service differentiation and QoS guarantee for heterogeneous traffic (e.g., video, voice and periodic scalar data), which is differentiated into three classes, i.e., real-time (RT) multimedia traffic, non-real-time (NRT) traffic and best effort (BE) traffic. Note that these DiffServ mechanisms are mainly proposed with one of or a hybrid of the following features: (i) adaptive contention window (CW) adjustment, (ii) dynamic duty cycling (changing active time), (iii) fragmentation and message passing for long packets, and (iv) intra-node and intra-queue prioritization [18]. However, these MAC protocols cannot fulfill the unique requirements of core-function-aware networks. Take NC-NET, for example; navigation service is the core function of NC-NET, which is essential to other multimedia services, even the services with stringent constraints on real-time variable bit rate, packet loss, and average delay. Thus, in this study, the design and analysis of the DiffServ mechanism, which can separate the kernel service from other functions, will be a basis for other technical details.

Cooperative communication is another effective technique for realizing the advantages of spatial diversity, especially when NC-NET requires robust and real-time data communication, or the communication is impacted by high mobility, intermittent connectivity, and unreliability of the wireless medium. Thus, the theory behind it, depending on the application scenarios, can be classified into two categories: cooperative diversity, and packet relay by selected neighbor nodes. Cooperative diversity aims to offset the multi-path fading effect of wireless channels through multiple cooperative antennas, which can maximize the total network channel capacity [19–21]. For instance, CoopMAC [20] takes full advantage of the broadcast nature of the wireless channel, thus creating spatial diversity. Packet relay by selected neighbor nodes improves the link utilization probability by transmitting through the selected relay nodes, instead of transmitting directly to the destinations. In previous works, EC-MAC in reference [22], and 2rcMAC in reference [23] are similar to our proposed protocol. EC-MAC adopts the best partnership selection algorithm to select the cooperative node with the properties of best channel conditions, highest transmission rate, and most balanced energy consumption. In 2rcMAC, the nodes update their relay table through a passive listening method, and obtain spatial diversity by a two-best-relay approach. However, this literature [20–23] is mainly confined to the scenarios with features such as short transmission range, low dynamic or static nodes, and omni-directional antennas. The cooperative MAC for large-scale sparse network is still an open issue, and does not have full-rate network codes for directional transmission [24–26]. Both of them restrain the use of cooperative communication in a practical large-scale dynamic network. Furthermore, in the majority of works, only the relays that can decode the packet successfully during the first transmission can be activated and get involved in the possible ARQ retransmission; meanwhile, the other relays keep silent during the whole ARQ process. Nevertheless, they ignore the fact that the relays, which cannot decode the packet correctly during the first transmission, still have the chance to decode the packet correctly in the ARQ process.

3. System Description and Related Models

The navigation-oriented NC-NET is described by an undirected graph $G = (C_g, E_g, R_g)$, where $C_g = \{C_1, \ldots, C_n\}$ denotes the set of sensor nodes and $C_i \in C_g$ means the node with a unique ID $i \in \{1, 2, \ldots, n\}$, $E_g \subseteq C_g \times C_g$ denotes the set of edge and edge $(C_i, C_j) \in E_g$ indicates that node $C_i \in C_g$ can receive the information from node $C_j \in C_g$, and $R_g = \{d_1, \ldots, d_m\}$ indicates the set of work radius and $d_x \in R_g$ means the work radius under the given data rate x Mbps.

3.1. Propagation and Channel Models

For simplicity, but without loss of generality, we ignore the unknown and heterogeneous factors of the signal propagation environment, and adopt a free space propagation model [27]:

$$P_{R(s \to r)} = G \frac{P_T}{d^\lambda}, 2 \le \lambda \le 4 \tag{1}$$

where $P_{R(s \to r)}$ denotes the received power of the packet from node s to r, P_T denotes the transmission power, d denotes the distance between transmitter and receiver, λ is a path loss exponent related to the propagation environment. $G = G_T G_R$, with G_T and G_R are antenna gains of transmitter and receiver. Thus, we have the absorption loss (in dB) $PL_{dB} = 92.44 + 10\lambda \log(f_{GHz}) + 10\lambda \log(d_{km})$, where PL_{dB} denotes the free space loss in dB, f_{GHz} is the electromagnetic frequency in GHz, and d_{km} is the distance between transmitter and receiver in km [28]. Let $f_{GHz} = 1$ GHz and $\lambda = 2$, combine PL_{dB} with the practical test results of the TTNT data link for the data rate [8], and then the propagation parameters can be summarized as Table 1.

Table 1. Propagation parameters in long-range transmission.

Free Space Losses (dB)	147.33	143.80	141.32	137.78
Frequency (GHz)	1	1	1	1
Distance (km)	555	370	278	185
Data Rate	220 kbit/s	500 kbit/s	1 Mbps	2 Mbit/s

For the sake of comparison with other protocols, let the propagation parameters (Table 1) combine with the channel assignment in IEEE 802.11b DCF [10]; the channel parameters can then be as shown in Table 2.

Table 2. Multi-rate setting for NC-NET ($f_{GHz} = 1$ GHz).

Serial No.	Data Rate	Threshold	Range
1	1 Mbit/s	−97 dBm	278 km
2	2 Mbit/s	−94 dBm	185 km
3	5.5 Mbit/s	−88 dBm	92.5 km
4	11 Mbit/s	−83 dBm	50 km

3.2. Antenna Model and Communication Modes

Assumption 1. (*Adaptive antenna*): Each node is equipped with an adaptive array antenna system, which has two transmission modes in terms of omni-directional and directional modes, as shown in Figure 2. Moreover, the receiving mode of the antenna system is omni-directional.

Assumption 2. (*Directional beam*): Let the ideal model of the beam coverage region be a spherical sector $V_{ideal}(\theta)$, where $R_{dir}(\theta)$ denotes the maximal permission range of sender s_2 ($s_2 \in C_g$), θ is the beam-width of its main-lobe. Suppose that the main-lobe gain is isotropic, and the gains of side-lobes and back-lobes are ignored. Moreover, the direction of the main-lobe can be rotated adaptively around the three-dimensional detecting region.

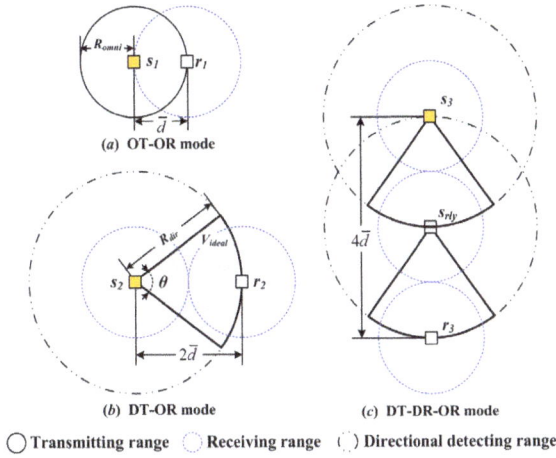

(a) OT-OR mode

(b) DT-OR mode   (c) DT-DR-OR mode

○ Transmitting range ◯ Receiving range ( ) Directional detecting range

Figure 2. Transmission modes with adaptive antenna.

The relationship between omni-directional transmission range R_{omni} and directional range $R_{dir}(\theta)$ can be expressed as follows:

Lemma 1. *Let the transmission be directional and the reception be omni-directional, the transmission range $R_{dir}(\theta)$ under the given transmission power P_T can be determined from*

$$R_{dir}(\theta) = \left(\frac{2\pi}{\theta}\right)^{\frac{1}{\lambda}} R_{omni} \qquad (2)$$

where λ is the path loss exponent, i.e., $2 \leq \lambda \leq 4$.

Let D be $(2\pi/\theta)^{1/\lambda}$ and under a given transmission power, the range of directional transmission (DT) is equivalent to D times of omni-directional transmission (OT). If an omni-directional broadcast request is received, or the azimuth of the destination node is unknown, the transmitter will send a packet in omni-directional mode (denoted by OT). If the azimuth of the destination node is known, or a directional transmission request is received, the transmission node will send a packet in directional mode (denoted by DT). Moreover, to extend the transmission radius, the relay node for retransmission mainly works in DT mode (denoted by DR). Thus, the communication modes can be specified as follows: (a) omni-directional transmission and omni-directional reception (OT-OR), (b) directional transmission and omni-directional reception (DT-OR), and (c) directional transmission, directional relay and omni-directional reception (DT-DR-OR). Let $d_{\mathrm{mod}-a}$, $d_{\mathrm{mod}-b}$, and $d_{\mathrm{mod}-c}$ be the ranges of the three modes, respectively, $D = 2$, and $R_{omni} = \bar{d}$; the maximal transmission range can be calculated by (2) as $d_{\mathrm{mod}-a} = \bar{d}$, $d_{\mathrm{mod}-b} = 2\bar{d}$, $d_{\mathrm{mod}-c} = 4\bar{d}$.

4. DiffServ-Based Dynamic Cooperative MAC Protocol

In this section, we introduce a DiffServ-based MAC protocol (DDC-MAC) and its key features for QoS-provisioning and adaptation according to different types of service. The proposed protocol consists of two compatible sub-protocols: (i) time-driven-based periodic communication, denoted as a PC sub-protocol, and (ii) event-driven-based on-demand communication, denoted as an OC sub-protocol.

4.1. MAC Sub-Protocol for Periodic Communication

In NC-NET, PC service is the core transmission demand with the characters in terms of delay sensitivity, packet loss sensitivity, and high communication frequency. In this section, we propose a collision-free MAC sub-protocol, which adopts DiffServ-based clustering topology and adapt-TDMA, to handle these periodic communication requests.

4.1.1. Adaptive TDMA Schedule

The basic hypothesis applicable to DiffServ-based clustering topology is outlined as follows:

Assumption 3. Under PC mode, only one member of a 1-hop cluster is allowed to send the packet at the same time, while the other nodes overhear the medium.

Assumption 4. The time axis of the adapt-TDMA schemes is divided by means of *round*, *frame*, and *mini-slot*, with the relationships *round* = $m \cdot frames$, $frame = k \cdot minislots$.

Assumption 5. The major function of a cluster-head C_{CH} is assumed to have the following features: (i) its function is to monitor, supervise, and gather the state change from its cluster members; (ii) it broadcasts a common management packet (denoted by CM) periodically, including a synchronized clock, access schedule (AS), etc.; and (iii) it is not designated as fixed relay nodes, in order to avoid the problems of "hot spots" and imbalance link loads.

As depicted in Figure 2, *round* indicates the total time for all *m* members of a cluster to update the data according to AS. *Frame* is the duration for one member to update its data, and can be subdivided into two *sub-frames*: a control sub-frame (*C-frame*) and a data sub-frame (*D-frame*). *C-frame* is the time for schedule, reservation, and contention window; *D-frame* is the time duration to handle the packet. *Mini-slot* is the fixed time duration to process the data less than l_x bits. $C - frame = k_1 \cdot minislots$, $D - frame = k_2 \cdot minislots$; herein, k_1 and k_2 are two known constants, which represent the lengths of the $C - frame$ and $D - frame$, respectively.

Furthermore, a parity field (PF) is introduced in the head of the *C-frame*, which is a 2-bit code with settings as follows: PF = 00 (i.e., overhear or idle mode in PC), PF = 01 (i.e., active mode in PC), PF = 10 (i.e., initial or idle mode in OC), PF = 11 (i.e., active mode in OC).

4.1.2. PC Sub-Protocol Based on Adapt-TDMA

This subsection elaborates the working mechanism for the PC sub-protocol, which includes the procedure of PC demand based on the adapt-TDMA mechanism and the random access mechanism for OC demand; additionally, a basic mechanism for the integration of PC service and OC service is introduced in a concise way. Figure 3 illustrates the schematic diagram of the PC sub-protocol, and the details are outlined as follows:

- **Step 1:** At the beginning of each *round*, cluster head C_{CH} broadcasts a CM packet by OT-OR mode, with the functions of correcting the synchronized clock and adjusting the access schedule (AS) according to the cluster member states of the recent *round*.
- **Step 2:** Cluster members reset their own respective AS according to the above new AS.
- **Step 3:** Cluster member C_i, which is at its sending *frame*, broadcasts a periodic exchange (PE) packet to its 1-hop neighbors by OT-OR mode; meanwhile, its PF is set to "01".
- **Step 4:** Other members $C_j (j \in (1, \cdots, m) \cap (j \neq i))$, which are in idle *frame*, receive the PE packet, and buffer the azimuth, localization, CSI information from C_i, and utilize this information for information fusion. The members in idle *frames* set the PFs to "00".
- **Step 5:** When every member has sent a PE packet according to their AS, one complete *round* ends, and then all the members resume *Step 1* for the next *round*.

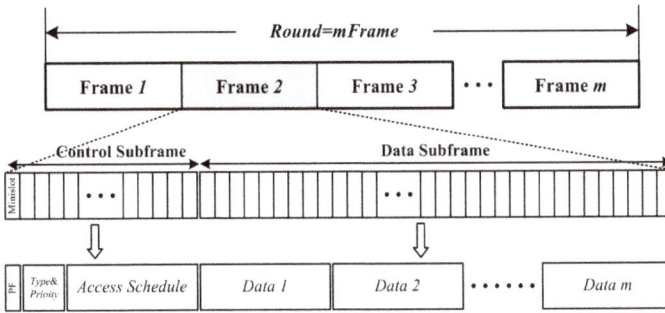

Figure 3. *Round* structure in adaptive TDMA schedule.

Moreover, if a random OC request has been sent to the intended receiver C_k after Step 2, the receiver C_k will implement the following procedures:

- If the PF of C_k is "00" without other OC requests, node C_k will receive and decode the RTS, and record the related data in its AS. Subsequently, node C_k will reply the request based on the procedures in Section 4.2, and then set its PF value as "10".
- If the PF of C_k is "01" and the channel of C_k is busy, node C_k will stay in PC mode, and the senders will compete for the access right of receiver C_k after a random back-off.
- If the PF of C_k is "00" with other OC requests, the multiple OC requests may be in contention at this timeslot, and thus compete for the access right of C_k.
- If node C_k is in OC mode and the DATA is transmitting, the PF of C_k will be set to "11". When this OC request is finished, its PF value will reset to "00", and then node C_k can resume the PC procedure.

Through the adapt-TDMA based sub-protocol above, the MAC of periodic communication services can be realized in a contention-free way. Thus, the OC request becomes the primary reason for channel contention and collision.

4.2. MAC Sub-Protocol for on-Demand Communication

The proposed on-demand MAC sub-protocol is mainly based on a CSMA/CA policy, a multi-channel structure, and a DiffServ mechanism. We first introduced a novel multiple priority mechanism, which is a critical part of DiffServ mechanisms.

Definition 1. *Service Priority, denoted by Pri.Si ($i \in \{1, 2, 3\}$), is defined as the allocation according to QoS requirements and tolerability in terms of average delay (AD) and packet loss (PL). Pri.S1 denotes AD sensitive and PL sensitive services (e.g., tracking data, video). Pri.S2 denotes AD tolerant and PL sensitive service (e.g., command data, and voice). Pri.S3 denotes AD tolerant and PL tolerant service. Note that Pri.S1 > Pri.S2 > Pri.S3.*

Definition 2. *Packet Priority, denoted by Pri.Pj, ($j \in \{1, 2, 3\}$), is the allocation based on the order of packet types in data transmission. Pri.P1 denotes DATA/ACK packet, Pri.P2 denotes CTS packet, Pri.P3 denotes RTS packet. Note that Pri.P1 > Pri.P2 > Pri.P3.*

The multiple priority mechanism is realized by encoding the related parity bit in packets and setting a lower threshold in CW size for higher priority traffic. The OC sub-protocol mainly consists of two phases, namely the Negotiation phase and Data transmission phase, to satisfy the OC access requests in a variety of scenarios.

4.2.1. Negotiation Phase

The negotiation phase aims to handle the pairs of three scenarios. **Scenario 1** is intra-cluster transmission, where both the sender and receiver are members of the same cluster; additionally, the pair pre-stores their prior knowledge of each other by PE packet (e.g., CSI, ID, and azimuth). **Scenario 2** is inter-cluster single-hop transmission, where the receiver and sender are not located at the same cluster, but are located in direct range of each other. **Scenario 3** is inter-cluster two-hop transmission, the receiver is located between one-hop and two-hop range of the sender. These scenarios will be our basis for the negotiation policy.

First, a rule for channel switch sequence, namely, adaptive channel back-off sequence (ACBS), will be encoded in RTS and CTS. The proposed ACBS is to detect and eliminate the threats in congested network. The ACBS is generated by the negotiation of the sender and the intended receiver based on prior knowledge, e.g., channels' idle probability, ID, and azimuth. When encountering collision among pairs, ACBS can act as a reference for the lower prior pair to switch among data channels directly without additional negotiation packets, thus effectively curtailing the process and improving the utilization rate of the medium. The negotiation phase of Scenarios 1 and 2 is depicted in Figure 4, with the details as follows:

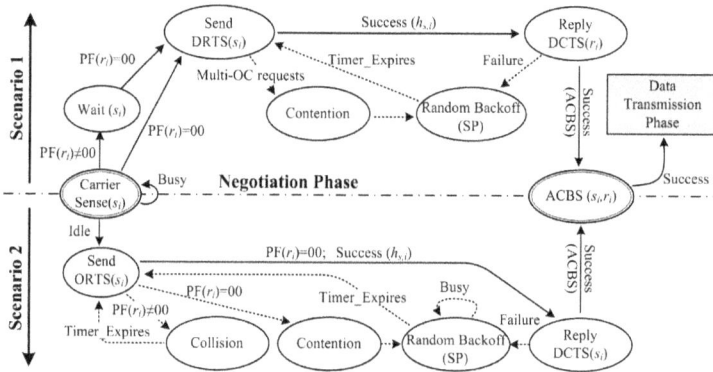

Figure 4. State diagram of Scenarios 1 and 2 in the negotiation phase.

Suppose that the sender s_i has an OC request to the intended receiver r_i, sender s_i will first sense the medium. If the medium is busy, sender s_i will postpone the negotiation and keep sensing. If some channel is idle, sender s_i will arrange the idle probability of the channels according to prior knowledge during the carrier sense, denoted by $h_{s,i}$, encode it into RTS, and then initiate the negotiation phase.

In Scenario 1, sender s_i has prior knowledge of receiver r_i before sending RTS, and thus can select an idle channel, and sends a RTS to receiver r_i through the *idle frame* of both s_i and r_i. If the PF of r_i is "00" and without other OC requests, r_i will receive the RTS, and decode the information of s_i, i.e., the service's type, data packet size, ID, and $h_{s,i}$. Thereafter, this information and the prior knowledge of receiver r_i (denoted by $h_{d,i}$) will be used to generate an ACBS for the pair (s_i, r_i), encoded in CTS, and then reply to s_i. Both RTS and CTS are sent in DT-OR mode. If the PF of r_i is "00" but with other OC requests, receiver r_i will sense the collision, and broadcast a warning packet, namely WARN, in order to prevent the sender from transmitting the above packet and notifying sender to switch channel immediately. Then, the senders will postpone randomly according to service priorities, and compete for the right of channel access again.

In Scenario 2, sender s_i has no prior knowledge of the intended receiver r_i except for its ID. Then s_i sends a RTS in OT-OR mode. If PF of r_i is "00" when RTS reaches it, the pair (s_i, r_i) then performs the same procedure as in Scenario 1. If PF of r_i is "01/11", r_i will send a WARN packet, sender s_i will postpone for a duration according to WARN, and will compete for the channel access rights again.

If sender s_i does not receive CTS after a time threshold, then it switches to the next data channel according to $h_{s,i}$, and sends RTS again.

4.2.2. Data Transmission Phase

The pair (s_i, r_i), having completed the negotiation phase, initiates the data transmission phase, and transmits packets by a data channel according to ACBS. In normal scenarios, sender s_i sends a DATA packet to receiver r_i, and then receiver r_i replies with an ACK packet to sender s_i after a time duration of SIFS when the OC procedure is completed. Note that both the DATA and ACK are transmitted in DT-OR mode, and both the sender and the receiver set their parity fields as "11", in order to avoid the disturbance from other pairs in the same data channel. The schematic diagram of data transmission phase is presented in Figure 5.

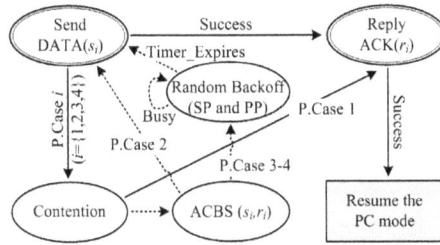

Figure 5. Schematic diagram of data transmission phase.

Note that some problematic scenarios might occur in this phase. *P.Case 1*: several pairs with different priorities select the same data channel, whereas pair (s_i, r_i) has the highest priority (SP and PP). *P.Case 2*: several pairs of different priorities select the same data channel, whereas pair (s_i, r_i) does not have the highest priority. *P.Case 3*: several pairs of the same priority select the same data channel. *P.Case 4*: the number of pairs, which intend to communicate, exceeds the number of available data channels.

The above scenarios will first be solved by the first-come-stay policy, i.e., the pair that switches to the data channel earliest will occupy the data channel; meanwhile, other co-existing pairs should switch to other data channels according to their ACBS. However, if several pairs with different priorities select the same data channel, only the pair with the highest priority can stay. Moreover, if several pairs with the same priority select the same data channel, all the collision pairs switch to the next data channel according to their ACBS.

4.3. MAC Sub-Protocol with Optimal Relay

This subsection aims to handle the MAC for the pairs in Scenario 3. It also aims to tackle the link failure problem caused by long-range transmission and degrading link quality.

4.3.1. Optimal Relay Selection Algorithm

Definition 3 (Optimal relay). *Given the pair (s, r), $s, r \in C_g$, its optimal relay s_{rly} is the node satisfying the following constraints. (i) Relay s_{rly} is within neighbor sets of nodes s and r, simultaneously. (ii) Relay s_{rly} can decode RTS and CTS of the pair (s, r) correctly. (iii) Relay s_{rly} holds the maximum SNR (denoted by γ_{SRD}) among the candidate relay set (CRS).*

Conditions (i) and (ii) are the prerequisites for a node to be a CRS member, and condition (ii) can be realized by checking the CRC of the received packet. To reduce the decoding overhead, we adopted an SNR-based checking method, in which a packet will be decoded only if the value of its instantly receiving SNR exceeds the SNR threshold; thus, the packet below the SNR threshold will be discarded. The following algorithms are designed to judge condition (iii).

Proposition 1 (Optimal link). *Given the quasi-stationary fading channel with free space channel amplitude, the link i_{max} with the maximum SNR value can be derived as follows:*

$$i_{max} = \underset{1 \leq i \leq n}{\mathrm{argmax}} \left\{ \gamma_{SRD}^i \right\}, i_{max} \in \{1, \ldots, n\} \tag{3}$$

where γ_{SRD}^i is the instantly receiving SNR value of the link i

$$\gamma_{SRD}^i = \frac{\gamma_{SR}^i \gamma_{DR}^i}{\gamma_{SR}^i + \gamma_{DR}^i + 1}, i \in \{1, \ldots, n\} \tag{4}$$

γ_{SR}^i *and* γ_{DR}^i *are the average receiving SNR from sender s to candidate relay s_{rly}^i and from s_{rly}^i to receiver r, respectively. As inspired by the literature [29,30], we design a back-off timer, in which the candidate relay starts its back-off timer as follows:*

Proposition 2 (Back-off timer). *Let the back-off time $T_{back-off}^{(i)}$ of the candidate relay s_{rly}^i be inversely proportional with the link quality γ_{SRD}^i, then s_{rly}^i sets its timer according to (5)*

$$T_{back-off}^{(i)} = \left[\frac{\gamma_{low}}{\gamma_{SRD}^i} \frac{T_{DIFS} - T_{SIFS}}{T_{slottime}} \right], i = 1, 2, \cdots, n \tag{5}$$

where $T_{back-off}^{(i)}$ is the back-off time of relay s_{rly}^i, γ_{low} is the threshold of γ_{SRD}^i for the relay s_{rly}^i intending to participate in retransmission, and n is the number of candidate relay within CRS.

The optimal relay, which has the optimal link quality and satisfies the constraints of Definition 3, will have the shortest back-off time and get access to the channel first.

4.3.2. Optimal Relay MAC Procedure

Suppose that the procedure for optimal relay initiates only if the direct transmission fails. First, we introduce two new control packets, namely Request to Cooperation (RTC) and Clear to Cooperation (CTC), with the same structure as RTS and CTS. The procedure can be elaborated as follows:

- **Step 1:** Source node s sends an encoded RTC & RTS packet to its 1-hop neighbors, this packet is encoded by RTC and RTS packets, and has the same structure as them. If node s knows the azimuth of node r based on prior knowledge, the packet will be sent in DT-OR mode; if not, in OT-OR mode with the basic data rate.
- **Step 2:** The 1-hop neighbors, which meet conditions (i) and (ii) of Definition 3, will decode the RTC & RTS packet, calculate γ_{SR}^i, and transmit RTS to node r after a SIFS.
- **Step 3:** Receiver r will decode the azimuth and position of node s after receiving RTS, and reply with CTS in DT-OR mode. The neighbors that satisfy conditions (i) and (ii) will be included into CRS. CRS members s_{rly}^i will calculate their γ_{SRD}^i by *Proposition 1*, and start the back-off timers. Hereafter, the optimal relay s_{rly} will access to the channel first and forward an encoded CTC & CTS packet to the source node. The CTC & CTS packet is encoded by CTC and CTS packets and has the same structure as them.
- **Step 4:** If the other relay candidates of CRS overhear the packet sent by s_{rly}, they will freeze their timer and keep on listening to the channels.
- **Step 5:** If the source node s receives CTC & CTS correctly, it will decode the information with respect to its destination r and the CRS. Then the pair (s, r) gets into the data transmission phase

with the assistant of the optimal relay s_{rly}, while other CRS members will delete the CTS after overhearing DATA, and then keep on listening to the channels.

- **Step 6:** If the other relay candidates do not overhear DATA after a SIFS, the cooperative communication fails for one time. Thus, the other relay candidates will reactivate their timers simultaneously. At this time, the suboptimal candidate, which has the channel quality next only to s_{rly}, will expire its timer first, and will execute Steps 3–5.

- **Step 7:** If no node leads to successful negotiation phase, or the number of retransmission attempts reaches the retry limit, the source node will then obtain channel access rights again for another *round* of packet transmission, and carry out the same procedures above.

4.4. Problematic Scenarios in DDC-MAC

4.4.1. Deafness and Hidden Terminal Problem

Figure 6a illustrates a scenario of the hidden terminal problem due to the use of pure omni-directional antennas (i.e., s_1, s_2, r_1, and r_2), and the deafness problem caused by using directional antennas (i.e., node s_3, and r_3). Suppose that sender s_2 intends to communicate with receiver r_2; meanwhile, the channel of receiver r_2 has been occupied due to overhearing the RTS packet from s_1, so the RTS from s_2 cannot be received by r_2, hence the hidden terminal problem occurs. Additionally, the deafness problem occurs when sender s_3 sends RTS through a pure directional antenna to the intended receiver r_3, whose antenna beam is pointing in a different direction. As a result, node r_3 is deaf to the RTS from sender s_3.

To address the two problems, in Scenario 1, the pairs (s_1, r_1) and (s_2, r_2) are equipped with adaptive antennas and communicate in DT-OR mode; thus, the two pairs can realize parallel transmission without interference. In Scenario 2, when node r_2 overhears the RTS from s_1, node s_2 will switch to another channel according to ACBS mechanism and send the RTS again. In addition, the deafness problem can also be resolved naturally through the adaptive antenna of r_3, which can receive the packet omni-directionally at all times. Then, node r_3 points its beam in the direction of s_3 shortly after receiving the RTS.

4.4.2. Interference and Congestion Problem

Figure 6c exhibits a problematic scenario for two reasons: (i) the distribution density of pairs is larger than network capacity; and (ii) the number of pairs exceeds the number of available data channels, thus the pairs might interfere with each other and result in network congestion. Figure 6d shows our proposed solutions, which can be expatiated as follows:

Case 1: *The receiver overhears packets from the direction of its sender.* For instance, the receiver r_5 of the pair (s_5, r_5) overhears the RTS or CTS from the pair (s_4, r_4) with a higher priority. Node r_5 will reset its NAV according to the overheard packet and copy the NAV value to the duration field of the WARN, which is used to notify sender s_5 to wait for the same duration. Then, sender s_5 will compete for the access right again after the time duration.

Case 2: *The sender overhears the packets from the direction of its receiver.* For instance, the sender s_6 of the pair (s_6, r_6) overhears a higher priority packet from the pair (s_3, r_3) during waiting for CTS. The sender s_6 will switch to another channel according to its ACBS immediately, and then receiver r_6 switches to the same channel automatically according to its ACBS after the duration of a timer, and competes for the access rights of the new channel again.

Case 3: *Channel congestion.* If a pair performs Case 1 and Case 2, but is still unable to communicate after switching among all data channels, for example, receiver r_1 of the pair (s_1, r_1) overhears the higher priority (PP or SP) packets simultaneously from the pairs (s_2, r_2), (s_3, r_3), and (s_4, r_4). In our solution, the pair (s_1, r_1) will switch to the first data channel of its ACBS, and wait for channel access rights until a data channel is idle, as shown in Figure 6d.

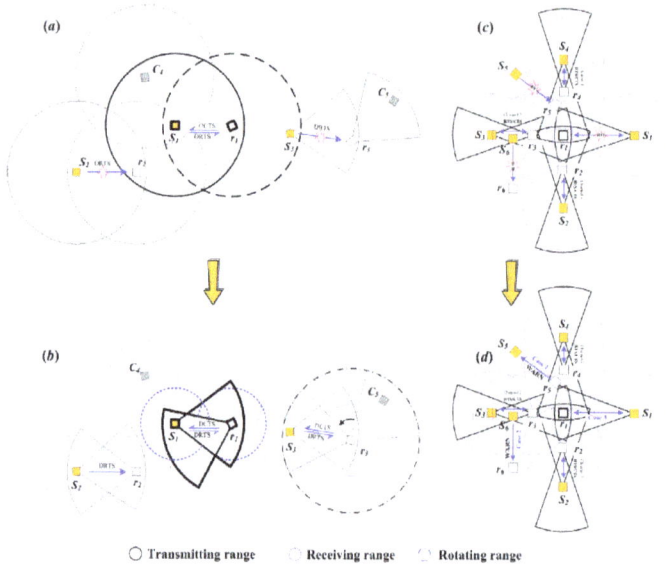

○ Transmitting range ○ Receiving range ○ Rotating range

Figure 6. Potential problematic scenarios in DDC-MAC. (**a**) problematic scenario of deafness and hidden terminal problem; (**b**) solutions of deafness and hidden terminal problem by DDC-MAC; (**c**) problematic scenario of interference and congestion problem; (**d**) solutions of interference and congestion problem by DDC-MAC.

5. Performance Analysis

This section analyzes the searching probability and throughput performances of DDC-MAC. Note that PC mode can proceed without collision, and thus can achieve an approximately 100% packet delivery ratio (PDR), and relatively fixed throughput. Thus, the analysis is mainly for OC mode, with the assumptions that: (i) the nodes are distributed uniformly in a 3D cluster network, and (ii) all senders are fully backlogged, and the size of the data payload is fixed.

5.1. Searching Probability of Next-Hop Nodes

In OC mode, the throughput has closed correlations with the location of next-hop nodes [21]. Thus, we should analyze the probability of searching next-hop nodes first. This probability is deduced in single-hop (Scenarios 1 and 2) and 2-hop (Scenario 3), respectively.

Lemma 2. *Given node $C_i^{(j)}$ ($j = 0, 1$), the next-hop relay of $C_i^{(j)}$ should satisfy two constraints: (i) node $C_i^{(j+1)}$ locates in three-dimensional sensing region of $C_i^{(j)}$ when the data rate is $x \in \{1, 2, 5.5, 11\}$ Mbps, (ii) the PDR is affected by fading effects and interference.*

In Scenarios 1 and 2, node $C_i^{(j+1)}$ is within the transmission range of node $C_i^{(j)}$ when the data rate is x Mbps; thus, node $C_i^{(j)}$ finds it easy to find node $C_i^{(j+1)}$ in DT-OR mode (Scenario 1) or OT-OR mode (Scenario 2). The node $C_i^{(j+1)}$ will decode the azimuth and localization of $C_i^{(j)}$ from the RTS packet, and reply CTS to $C_i^{(j)}$ with a delivery ratio of approximately 100%, thus the next-hop searching probability in Scenarios 1 and 2 can be derived as:

$$P_{1-hop}(x,y,l) = \begin{cases} 1 - \beta, & l \in [l_{threshold}, \bar{d}), \text{ Scenario 1} \\ (1 - \beta)V_{dir}(d_x, \theta_{max})/V_{omni}(d_x), & l \in [l_{threshold}, 2\bar{d}), \text{ Scenario 2} \end{cases} \tag{6}$$

where $l_{threshold}$ is the minimum unperturbed distance between $C_i^{(j)}$ and $C_i^{(j+1)}$, $V_{dir}(d_x, \theta_{max})$ is the coverage region of directional beam of $C_i^{(j)}$ when its beamwidth is θ_{max} and transmission range is d_x, β is the fading effects and interference on PDR, $V_{omni}(x)$ is the detection region of directional beams (with data rate x), which can be expressed as follows:

$$V_{dir}(d_x, \theta_{max}) = \frac{2}{3}\pi d_x^3 \left[1 - \cos\frac{\theta_{max}}{2}\right], \quad V_{omni}(d_x) = \frac{4}{3}\pi d_x^3 \tag{7}$$

In Scenario 3, receiver $C_i^{(j+2)}$ exceeds the direct range of its sender $C_i^{(j)}$, so only the node $C_i^{(j+1)}$, which is located in the ranges of both $C_i^{(j)}$ (with data rate x) and $C_i^{(j+2)}$ (with data rate y), can be used as a next-hop node. In other words, $C_i^{(j+1)}$ should be located in the overlapping region of $C_i^{(j)}$ and $C_i^{(j+2)}$ when the distance is l km ($l \leq d_x + d_y$). The transmission mode of Scenario 3 includes DT-DR-OR mode and OT-OR-OR mode, as shown in Figure 7, thus only the node $C_i^{(j+1)}$ within the hatch area can meet the conditions in Lemma 2.

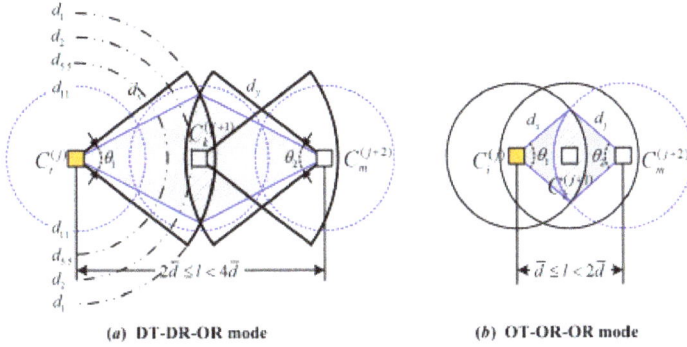

(a) DT-DR-OR mode   (b) OT-OR-OR mode

Figure 7. Searching probability region in Scenario 3.

Suppose that the distance between $C_i^{(j)}$ (with data rate x) and $C_i^{(j+2)}$ (with data rate y) is l km ($l \leq d_x + d_y$), the volume of the hatch area $V_0(d_x, d_y, l)$ in Figure 7 can be calculated as

$$V_0(d_x, d_y, l) = \frac{\pi}{6}d_x^3 \cdot \left[4 - 3\sin^3\frac{\theta_1}{2} - \cos^3\frac{\theta_1}{2} + 3\cos\frac{\theta_1}{2}\right] + \frac{\pi}{6}d_y^3 \cdot \left[4 - 3\sin^3\frac{\theta_2}{2} - \cos^3\frac{\theta_2}{2} + 3\cos\frac{\theta_2}{2}\right] \tag{8}$$

where θ_1 is beamwidth angle of node $C_i^{(j)}$, θ_2 is beamwidth angle of $C_m^{(j+2)}$. In DT-DR-OR mode, $\theta_1 \in [0, \theta_{max}]$ and $\theta_2 \in [0, \theta_{max}]$, whereas in OT-OR-OR mode, $\theta_1 \in [0, \pi)$ and $\theta_2 \in [0, \pi)$.

When the distance between $C_i^{(j)}$ (with data rate x) and $C_i^{(j+2)}$ (with data rate y) is $l \in [\bar{d}, 2\bar{d}]$ km, the searching probability for $C_i^{(j+1)}$ in OT-OR-OR mode can be calculated as

$$P_{2hop\_1}(d_x, d_y, l) = \begin{cases} (1-\beta)\frac{V_0(d_x,d_y,l)}{V_{omni}(d_x)}, & \text{Condition I} \\ (1-\beta)\frac{V_0(d_x,d_y,l)}{V_{omni}(d_x)} + \beta\frac{V_0(d'_x,d'_y,l)}{V_{omni}(d'_x)}, & \text{Condition II} \end{cases} \tag{9}$$

where d'_x and d'_y denote the transmission ranges of a pair with the priority (i.e., SP or PP) higher than that of $C_i^{(j)}$ and $C_i^{(j+2)}$, *Condition I* denotes that pair ($C_i^{(j)}, C_i^{(j+2)}$) has the highest priority, or that contestant pairs with higher priority do not exist. *Condition II* means that the contestant pairs are of higher priority than pair ($C_i^{(j)}, C_i^{(j+2)}$).

In addition, the searching probability for $C_i^{(j+1)}$ in DT-DR-OR mode is affected by the beam direction due to the adoption of directional antenna. Let the influence of beam direction on searching probability be ξ, the searching probability for $C_i^{(j+1)}$ can be deduced as

$$P_{2hop\_2}(d_x, d_y, l) = \begin{cases} (1 - \xi)(1 - \beta) \frac{V_0(d_x, d_y, l)}{V_{dir}(d_x)}, & \text{Condition I} \\ (1 - \xi) \left[(1 - \beta) \frac{V_0(d_x, d_y, l)}{V_{dir}(d_x, \theta_{max})} + \beta \frac{V_0(d_{x'}, d_{y'}, l)}{V_{dir}(d_{x'}, \theta_{max})} \right], & \text{Condition II} \end{cases} \tag{10}$$

where $l \in [2\bar{d}, 4\bar{d})$. In Scenarios 1–3, among n neighbor nodes, there exists at least one such node, with the probability as follows:

$$P'(d_x, d_y, l) = 1 - \sum_{(x',y')>(x,y)} P'(d_{x'}, d_{y'}, l) - \left(1 - \sum_{(x',y')\geq(x,y)} P(d_{x'}, d_{y'}, l) \right)^n \tag{11}$$

The searching probability of next-hop nodes in single-hop and 2-hop transmissions has been deduced as described above; thus, the probability $P'(d_x, d_y, l)$, which is closely related to the data rate $(x, y \in \{1, 2, 5.5, 11\})$ and other influences (ξ and β), will play a critical role in the analysis of saturated throughput.

5.2. Saturated Throughput Analysis

The OC mode in Scenarios 1 and 2 is within the direct transmission range, and its negotiation is partly based on the CSMA/CA in IEEE 802.11b DCF [6,28]. Thus, the per-hop throughput can be deduced by Bianchi's Equation [29]

$$S_{1-hop} = \frac{P_s P_{tr} E[p]}{(1 - P_{tr})\sigma + P_{tr} P_s T_s + P_{tr}(1 - P_s)T_c} \tag{12}$$

where P_{tr} denotes the probability that at least one node transmitting an RTS in a given time slot; P_s denotes the probability that one node transmitting an RTS in the given time slot; $E[p]$ is the average packet payload size, T_c is the average time that the channel is sensed busy by each node during a collision. T_s represents the expected transmission time that a node transmits a packet without collision; due to the analysis of Scenarios 1 and 2 in Section 4.2, T_s can be calculated as

$$T_s = \sum_{x \in \{1,2,5.5,11\}} P_x T_{dir}(x) \tag{13}$$

where P_x is the probability for the data rate between $C_i^{(j)}$ and $C_i^{(j+2)}$ adopting x Mbps,

$$P_{11} = \frac{d_{11}^3}{d_1^3}, P_{5.5} = \frac{(d_{5.5}^3 - d_{11}^3)}{d_1^3}, P_2 = \frac{(d_2^3 - d_{5.5}^3)}{d_1^3}, P_1 = \frac{(d_1^3 - d_2^3)}{d_1^3} \tag{14}$$

In Equation (13), $T_{dir}(x)$ is the time duration of successful transmission for the packets, including RTS, CTS, DATA, and ACK, with the data rate x Mbps, and can be calculated as follows

$$T_{dir}(x) = T_{rts} + T_{cts} + \frac{E[p]}{r_x} + T_{ack} + 3T_{sifs} + 4\sigma + T_{difs} \tag{15}$$

where σ is the propagation delay. During collision, each node senses that the channel is busy, and the collision time T_c can be deduced by

$$T_c = T_{rts} + \sigma + T_{difs} \tag{16}$$

The saturated throughput in Scenario 3 should take the overhead of cooperation and optimal relay selection into consideration. Then Equation (13) can be rewritten as follows

$$T_s = \sum_{x \in \{1,2,5.5,11\}} P_x T_{relay}(x) \tag{17}$$

where $T_{relay}(x)$ is the transmission time between the pair $(C_i^{(j)}, C_i^{(j+2)})$ with the transmission rate x Mbps, $T_{relay}(11) = 2 \cdot T_{dir}(11)$, $T_{relay}(5.5) = 2 \cdot T_{dir}(5.5)$. $T_{relay}(2)$ can be calculated from the searching probability for $C_i^{(j+1)}$ and the benefit by the relay node as follows:

$$T_{relay}(2) = T_{co\_oh}(P'(d_{11}, d_{11}, 2) + P'(d_{5.5}, d_{11}, 2) + P'(d_{5.5}, d_{5.5}, 2))$$
$$+2P'(d_{11}, d_{11}, 2)E[p]/d_{11} + P'_\alpha(d_{5.5}, d_{11}, 2) \cdot E[p]\left(d_{11}^{-1} + d_{5.5}^{-1}\right) + 2P'(d_{5.5}, d_{5.5}, 2)E[p]/d_{5.5} \tag{18}$$
$$+(1 - P'(d_{11}, d_{11}, 2) - P'(d_{5.5}, d_{11}, 2) - P'(d_{5.5}, d_{5.5}, 2))E[p]/d_2$$

where T_{co_oh} is the transmission time of a control message in optimal relay transmission.

$$T_{co\_oh} = T_{difs} + T_{rts} + T_{cts} + T_{ack} + 4T_{sifs} + T_{back-off} + T_{sifs} \tag{19}$$

Moreover, $T_{relay}(1)$ can be derived similarly, and the saturated throughput in Scenario 3 can be calculated with Equation (12) by substituting Equation (17) into it.

Through the deductions above, we complete the deduction of the saturated throughputs, which is applicable for the proposed DDC-MAC protocol under the three transmission scenarios concerned, and these results will be used as the basis for the calculation of saturated throughput in the following simulation-based analysis.

6. Performance Evaluation

In this section, we present simulation-based studies to evaluate the performance of the proposed protocol based on ns-2 simulator (version 2.29) and Matlab. We compare our protocol with four other well-known MAC protocols; IEEE 802.11b DCF [6] (dented by 802.11 DCF), LMAC [12], IEEE 802.11e EDCA [13] (denoted by 802.11 EDCA), DMAC [5] and CoopMAC [19]. In addition, the simulation setup, the results, and the analysis of different scenarios are presented.

6.1. Simulation Setup

In our simulations, 160 sensor nodes are distributed schematically in a cylindrical region with a radius of 250 km and an altitude of 20 km. The nodes are in a sparse setting, and are in clusters based on DMDG protocol. The initial topology is shown in Figure 8. The nodes are designated into three classes: low mobility nodes ($Group_1$1–2), with the velocity randomly varied between [0, 18] m/s; middle mobility nodes ($Group_2$1–3), with the velocity varied between [18, 100] m/s; and high mobility nodes ($Group_2$4–6), with the velocity varied between [100, 300] m/s. In each simulation, the percentage of each class of nodes varies from 10% to 30%, with an increment of 10%.

All nodes are identical, and use hybrid antenna arrays in free space ($\lambda = 3, \theta = \pi/4$); and all active nodes are saturated, i.e., their data buffers are always nonempty. The default parameters used in the simulations and the analyses are summarized in Table 3.

Furthermore, three kinds of simulation scenarios are conducted: (i) all nodes communicate in PC mode; (ii) parts of nodes communicate in 1-hop transmission range, that is, OC mode in Scenarios 1 and 2; (iii) parts of nodes communicate exceeding 1-hop transmission range, that is, OC mode in Scenario 3.

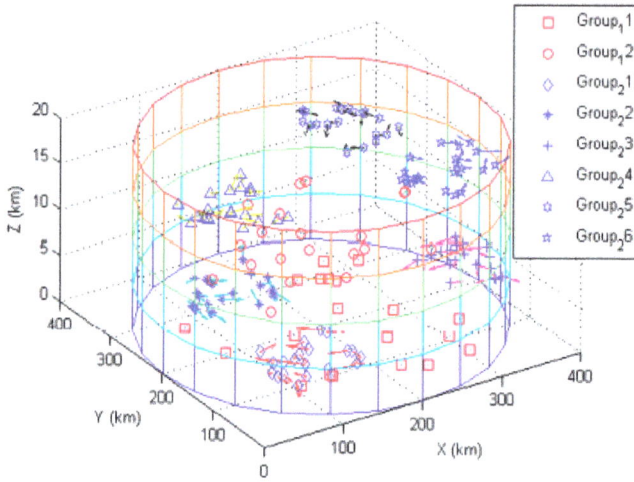

Figure 8. Initial network topologies with 160 nodes.

Table 3. Analytical parameter settings.

Parameters	Values	Parameters	Values
Data rate (Mbps)	{1, 2, 5.5, 11}	*CW for Pri.S3*	$CW_{min} = 32$, $CW_{max} = 1024$
SIFS	0.01 ms	ACK	14 bytes
DIFS	0.02 ms	DATA packet	512 bytes
ACK time	0.02 ms	MAC header	34 bytes
SNR threshold	15 dB	PHY header	24 bytes
RTS	20 bytes	Tx power (dBm)	32(omni), 38(dir)
CTS	14 bytes	Cluster Protocol	DMDG [2]
CW for Pri.S1	$CW_{min} = 8$, $CW_{max} = 16$	Routing Protocol	FMCQR [3]
CW for Pri.S2	$CW_{min} = 16$, $CW_{max} = 32$	—	—

6.2. Simulation Results

6.2.1. Saturated Throughput and Delay in Periodic Communication

This subsection evaluates the saturated throughput (ST) and average delay (AD) of DDC-MAC in PC scenario; the results are compared to the typical schedule-based MAC protocol (i.e., LMAC) and the typical contention-based MAC protocol (i.e., IEEE802.11b DCF, DMAC). The packet arrival rate (PAR) of each node is 20 *packets/sec*. To consider the impacts of fading effects and the influence of beam direction on saturation, the transmission rate was down-graded with the probabilities β and ξ. Thus, we set $\beta = 0.20$ and $\xi = 0.20$, and increase the number of nodes from 40 to 160; the analysis of saturated throughput can be validated by observing the aggregate uplink throughput.

Figure 9a compares the ST performance of four protocols with PC service. The results indicate that the throughputs of DDC-MAC, LMAC, and DMAC increase linearly with the increase of network size ($N_{node} \in [40, 160]$). The reason is that LMAC and DDC-MAC adopt conflict-free TDMA schedule, while DMAC adopts a simple duty-cycle-based operation. Meanwhile, 802.11 DCF obtains the worst throughput performance. When $N_{node} > 80$, the throughput of 802.11 DCF stays at a constant value ($Th_{dcf-pc} = 780$ kbps), and then decreases slowly when $N_{node} > 100$. The reason for this is that the increasing pairs aggravate the contentions and collisions, and the capacity for communications is in saturation, i.e., 570 kbps when $N_{node} = 160$. Furthermore, these results can validate the analysis of saturation, which is presented in Section 5.2.

Figure 9b depicts the average latencies under PC service. It is clear that the average delay of DDC-MAC, LMAC, and DMAC increases slowly when the network size is in $N_{node} \in [40, 160]$, whereas the performance of 802.11 fluctuates due to its omni-directional transmission and CSMA-based mechanism. In addition, LMAC and DDC-MAC achieve a similar inclination as DMAC, but their average delays are much lower, as they enjoy the benefits of a TDMA-based mechanism, thus guaranteeing non-collision transmission. Moreover, DDC-MAC outperforms LMAC because its traffic period is shortened by cluster-based topology, and its average delay is lower than 55 ms when $N_{node} = 160$.

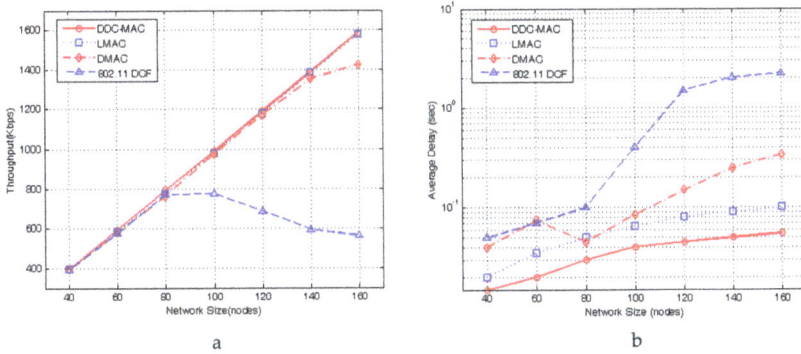

Figure 9. Performance of MAC protocols in PC mode: (**a**) Throughput; (**b**) End-to-end average delay.

6.2.2. Saturated Throughput and Delay in 1-Hop on-Demand Communication

This subsection evaluates the ST and AD performance of DDC-MAC in 1-hop OC mode under single-channel (SC) and multi-channel (MC) environments, respectively. We assume that each node in 802.11 DCF is equipped with an omni-directional antenna, and the node in 802.11 EDCA and DDC-MAC is equipped with the adaptive antenna array proposed in Section 3.2. To verify the proposed Diffserv mechanism, the results of different service priorities, namely Pri.S1, Pri.S2, and Pri.S3, are depicted in Figure 10. As shown in the results, in general, the service with higher service priority can achieve the performance of throughput and average delay superior to those of lower priority services (Pri.S1 > Pri.S2 > Pri.S3), whereas the performances in MC environments outperform those of SC environments.

Figure 10a,b depicts the ST comparison of three protocols under SC and MC environments. In Figure 10a, the throughput of 802.11 DCF increases slowly due to its OT transmission mode, and thus results in a high probability of contentions and collisions. Meanwhile, 802.11 EDCA and DDC-MAC explore the opportunities of spatial reuse effectively by using adaptive antenna array; thus, their throughputs increase along with the number of nodes. Finally, compared to 802.11 DCF, 802.11 EDCA and DDC-MAC achieve approximately 45% and 70% of throughput gain, respectively. When $N_{node} > 70$, the throughput growth of 802.11 EDCA and DDC-MAC decelerates and holds a constant value when $N_{node} = 100$ ($Th_{dmac-sc} = 7540$ kbps, $Th_{ddcmac-sc} = 8760$ kbps). In a SC environment, DDC-MAC cannot use ACBS for distributing communicating nodes over multiple channels; thus, its throughput performance is similar to 802.11 EDCA. Figure 10b depicts the ST of three protocols in an MC environment, in which 802.11 DCF and 802.11 EDCA use a random channel selection policy, and the proposed DDC-MAC uses an ACBS policy. The throughputs of three protocols are similar when the number of pairs is few ($N_{node} \leq 30$). However, when the number of pairs increases ($N_{node} > 30$), the throughputs of 802.11 EDCA and DDC-MAC are superior to that of 802.11 DCF because of the exploitation of spatial reuse opportunities by means of directional antenna and ACBS mechanisms. Moreover, DDC-MAC outperforms 802.11 EDCA because its traffic congestion is alleviated by the DMDG-based cluster protocol. The ST performance in an MC environment outperforms that of SC environment by 20% to 32%.

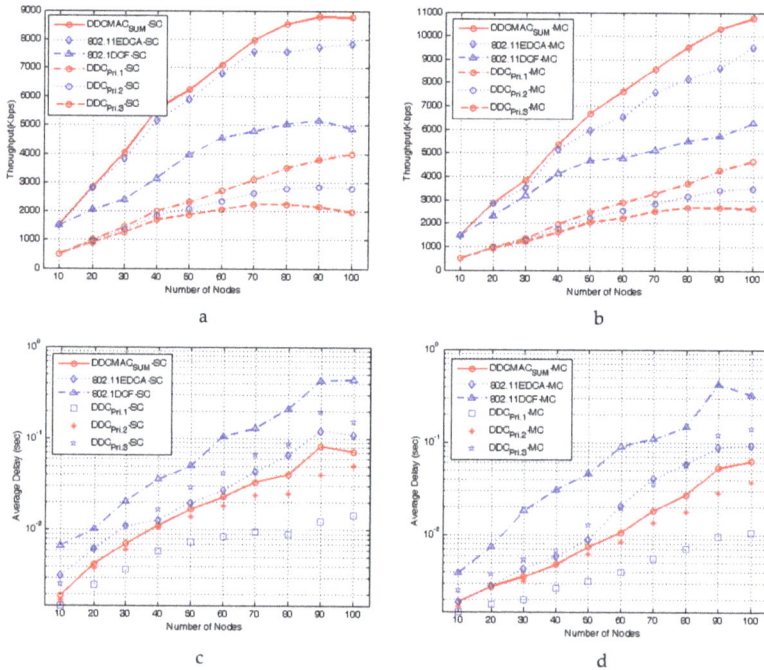

Figure 10. Throughput and average delay performances versus number of nodes in 1-hop OC mode. (**a**) Throughput in single-channel environment; (**b**) Throughput in multi-channel environment; (**c**) average delay in single-channel environment; (**d**) average delay in multi-channel environment.

Figure 10c,d describes the impact of the number of nodes on packet delay. The packet delay is measured by the duration from the memory buffer to the time when the packet has been sent successfully. Figure 10c shows the comparison of the three protocols in SC environment. Generally, the AD increases along with the number of nodes, owing to increasing of contentions and collisions. The 802.11 DCF adopts an omni-directional antenna, which blocks parallel communications, and thus its packet delay increases the fastest. Meanwhile, the AD of 802.11 EDCA and DDC-MAC is compensated for by exploiting spatial reuse opportunities by directional antennas. Compared to Figure 10c, the performance in multi-channel environments is superior in every respect, with the average delay decreasing by approximately 40–50%. As shown in Figure 10d, the ADs of three protocols are constant (about 2.5 ms) when $N_{node} \leq 30$. Since there are three data channels in use, the probability of collision and retransmission is low when N_{node} is small. However, the AD of 802.11 DCF increases rapidly with the increase of N_{node}, and reaches 0.42 s when $N_{node} = 100$. Meanwhile, the AD of 802.11 EDCA and DDC-MAC increases slower than 802.11 DCF, and keeps a constant value when $N_{node} = 100$ (62.3 ms for DDC-MAC, 92.6 ms for 802.11 EDCA).

6.2.3. Performance Analysis in 2-Hop Optimal Relay Communication

In this subsection, we present a numerical analysis for the performance with the increasing transmission range and the degrading link quality. The number of nodes is set to 70.

Figure 11a,b depicts the impact of transmission distance on bit error ratio (BER) and throughput performance. It is observed that 802.11 EDCA, CoopMAC, and DDC-MAC have similar BER and throughput when the distance is closer than 150 km. When the distance exceeds 150 km, the performance of 802.11 EDCA and 802.11 DCF degrades sharply due to the deterioration of link quality and retransmission. When the distance is 400 km, the BER of 802.11 EDCA and 802.11 DCF all keep constant values to 95%, and their throughputs keep constant values to 530 kbps, while CoopMAC and DDC-MAC performs better than 802.11 DCF and 802.11 EDCA due to the benefits of the cooperative relay mechanism. Furthermore, DDC-MAC outperforms CoopMAC due to the optimal relay mechanism and the ACBS mechanism. When the transmission range is 400 km, the bit error ratios of CoopMAC and DDC-MAC keep constant values of 38.5% and 29.7%, and the throughputs are 4750 kbps and 6430 kbps, respectively.

Figure 11c,d shows the PDR and AD performance of four protocols with environmental noise level varied from −94 dBm to −76 dBm. When the channel condition is good (*Noise < −90 dBm*), the PDR of four protocols is higher than 94%, the AD is lower than 0.1 s, and degrades slowly. When the noise level increases (*Noise > −90 dBm*), the performance of 802.11 EDCA and 802.11 DCF degrades sharply due to the deterioration of link quality. When *Noise = −75 dBm*, the PDR of 802.11 EDCA and 802.11 DCF is 21%, and the AD is 0.54 s and 0.72 s, respectively. CoopMAC and DDC-MAC outperforms 802.11 DCF and 802.11 EDCA due to the benefits from channel selection policy, and DDC-MAC outperforms CoopMAC due to the optimal relay mechanism and its adaptive antenna array. When *Noise = −75 dBm*, the PDRs of CoopMAC and DDC-MAC are 48.1% and 68.3%, and the AD is 0.33 s and 0.12 s, respectively.

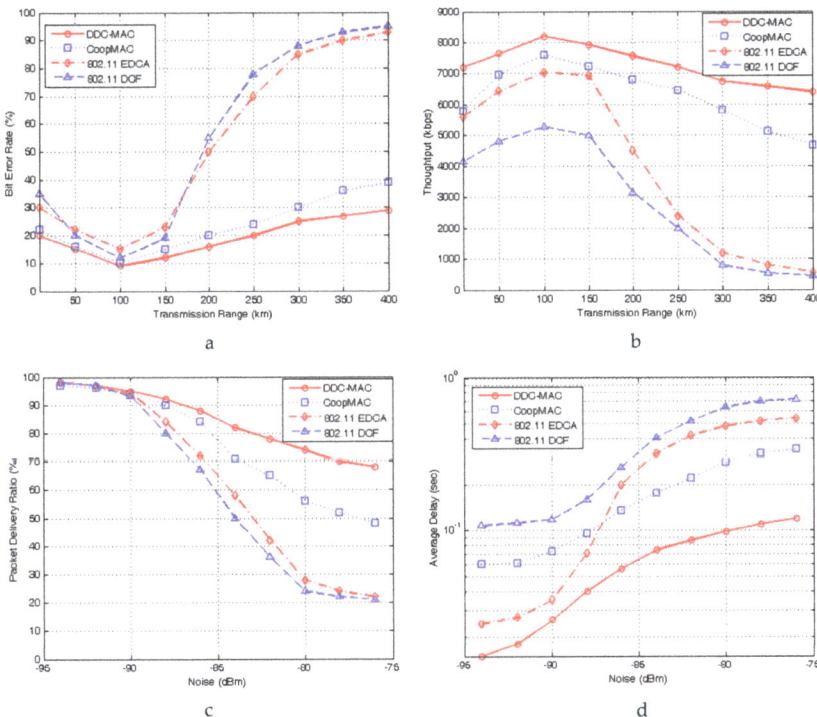

Figure 11. Impacts of transmission range and link quality on MAC performances with 2-hop optimal relay. (**a**) Transmission distance vs. bit error ratio; (**b**) transmission distance vs. throughput; (**c**) packet delivery ratio vs. link quality; (**d**) average delay vs. link quality.

7. Conclusions and Future Work

In this paper, we have proposed a novel approach, namely a DiffServ-based dynamic cooperative MAC (DDC-MAC) protocol, to realize QoS media access control for the wireless network with the unique features in terms of large-range transmission, high dynamic topology, three-dimensional monitor region, and heterogeneous services.

Its performance is analyzed by Markov chain-based modeling, and validated by ns-2 simulations. Both theoretical analysis and simulation experiments show that DDC-MAC can leverage cooperative communications and exploit spatial and user diversities. Therefore, it outperforms the existing protocols under the same channel assumptions and network scenarios. DDC-MAC also provides extended service ranges and a robust wireless communication link in NC-NET.

As future research, we will pursue the following directions: (i) investigate protocols integrated with routing protocols; (ii) explore the performance with different slot intervals, prediction window and prediction techniques, and (iii) extend DDC-MAC to other NC-NET with different topologies and mobility models.

Acknowledgments: This work was supported in part by the National Natural Science Foundation of China under Grant No. 61473306, and partly by the Defense Advanced Research Project of China under Grant No. 9140A09040614JB14001.

Author Contributions: Chao Gao and Guorong Zhao conceived and designed the MAC protocol; Bin Zeng analyzed and developed algorithm; Jianhua Lu performed the experiments; Chao Gao wrote the paper.

Conflicts of Interest: The authors declare no conflict of interest.

References

1. Bekmezci, I.; Sahingoz, O.; Temel, S. Flying Ad-Hoc Networks (FANETs): A survey. *Ad Hoc Netw.* **2013**, *11*, 1254–1270. [CrossRef]
2. Gao, C.; Zhao, G.; Pan, S. Distributed multi-weight data-gathering and aggregation protocol in fleet wireless sensor networks: Optimal and heuristic algorithms. *Int. J. Intell. Eng. Syst.* **2009**, *2*, 1–8. [CrossRef]
3. Gao, C.; Zhao, G.; Lu, J.; Pan, S. A Grid-based Cooperative QoS Routing Protocol with Fading Memory Optimization for Navigation Carrier Ad Hoc Networks. *Comput. Netw.* **2015**, *76*, 294–316. [CrossRef]
4. Gao, C.; Zhao, G.; Lu, J.; Pan, S. Decentralised Moving-Horizon State Estimation for a Class of Networked Spatial-Navigation Systems with Random Parametric Uncertainties and Communication Link Failures. *IET Control Theory Appl.* **2015**, *9*, 2666–2677. [CrossRef]
5. Ko, Y.; Shankarkumar, V.; Vaidya, N. Medium access control protocols using directional antennas in ad hoc networks. In Proceedings of the Nineteenth Annual Joint Conference of the IEEE Computer and Communications Societies, Tel Aviv, Israel, 26–30 March 2000; pp. 13–21.
6. IEEE 802.11 Working Group. Wireless LAN medium access control (MAC) and physical layer (PHY) specifications. In *ANSI/IEEE Standard 802.11*; IEEE 802.11 Working Group: Berlin, Germany, 1999.
7. Yi, S.; ·Pei, Y.; ·Kalyanaraman, S.; Azimi-Sadjadi, B. How is the capacity of ad hoc networks improved with directional antennas. *Wirel. Netw.* **2007**, *13*, 635–648. [CrossRef]
8. Collins, R. Tactical Targeting Network Technology (TTNT) Communicating at the Speed of Battle [EB/OL]. October 2010. Available online: http://www.rockwellcollins.com/content/pdf/pdf.7501.pdf (accessed on 15 October 2010).
9. Alam, M.; Hussain, M.; Kwak, K. Neighbor initiated approach for avoiding deaf and hidden node problems in directional MAC protocol for ad-hoc networks. *Wirel. Netw.* **2013**, *19*, 933–943. [CrossRef]
10. Sakakibara, K.; Taketsugu, J. A new IEEE 802.11 DCF utilizing freezing experiences in backoff interval and its saturation throughput. *J. Commun. Netw.* **2010**, *12*, 43–51.
11. Huang, Z.; Shen, C.; Srisathapornphat, C.; Jaikaeo, C. A busy-tone based directional MAC protocol for ad hoc networks. In Proceedings of the Military Communications Conference (MILCOM 2002), Anaheim, CA, USA, 7–10 October 2002; pp. 1233–1238.
12. Incel, O.; Hoesel, L.; Jansen, P.; Havinga, P. MC-LMAC: A multi-channel MAC protocol for wireless sensor networks. *Ad Hoc Netw.* **2011**, *9*, 73–94. [CrossRef]

13. Hoesel, L. Sensors on Speaking Terms: Schedule-Based Medium Access Control Protocols for Wireless Sensor Networks. Ph.D. Thesis, University of Twente, Enschede, The Netherlands, June 2007.

14. Akyildiz, I.; Melodia, T.; Chowdhury, K. A survey on wireless multimedia sensor networks. *Comput. Netw.* **2007**, *51*, 921–960. [CrossRef]

15. Lazarou, G.; Li, J.; Picone, J. A cluster-based power-efficient MAC scheme for event-driven sensing applications. *Ad Hoc Netw.* **2007**, *5*, 1017–1030. [CrossRef]

16. Boukerche, A.; Dash, T. Performance evaluation of a generalized hybrid TDMA/CDMA protocol for wireless multimedia with QoS adaptations. *Comput. Commun.* **2005**, *28*, 1468–1480. [CrossRef]

17. Saxena, N.; Roy, A.; Shin, J. Dynamic duty cycle and adaptive contention window based QoS-MAC protocol for wireless multimedia sensor networks. *Comput. Netw.* **2008**, *52*, 2532–2542. [CrossRef]

18. Yigitel, M.; Incel, O.; Ersoy, C. Design and implementation of a QoS-aware MAC protocol for Wireless Multimedia Sensor Networks. *Comput. Commun.* **2011**, *34*, 1991–2001. [CrossRef]

19. Liu, P.; Tao, Z.; NaNarayana, S.; Korakis, T.; Panwar, S. CoopMAC: A cooperative MAC for wireless LANs. *IEEE J. Sel. Areas Commun.* **2007**, *25*, 340–354. [CrossRef]

20. Moh, S.; Yu, C. A cooperative diversity-based robust MAC protocol in wireless ad hoc networks. *IEEE Trans. Parallel Distrib. Syst.* **2011**, *22*, 353–363. [CrossRef]

21. An, D.; Woo, H.; Yoon, H.; Yeom, I. Enhanced Cooperative Communication MAC for Mobile Wireless Networks. *Comput. Netw.* **2013**, *57*, 99–116. [CrossRef]

22. Zhang, X.; Guo, L.; Wei, X. An energy-balanced cooperative MAC protocol based on opportunistic relaying in MANETs. *Comput. Electr. Eng.* **2013**, *39*, 1894–1904. [CrossRef]

23. Khalid, M.; Wang, Y.; Ra, I.; Sankar, R. Two-Relay-Based Cooperative MAC Protocol for Wireless Ad hoc Networks. *IEEE Trans. Veh. Technol.* **2011**, *60*, 3361–3373. [CrossRef]

24. Antonopoulos, A.; Skianis, C.; Verikoukis, C. Network coding-based cooperative ARQ scheme for VANETs. *J. Netw. Comput. Appl.* **2013**, *36*, 1001–1007. [CrossRef]

25. Wang, X.; Zhang, X.; Chen, G.; Zhang, Q. Opportunistic cooperation in low duty cycle wireless sensor networks. In Proceedings of the IEEE International Conference on Communications (ICC), Cape Town, South Africa, 23–27 May 2010.

26. Xie, K.; He, S.; Zhang, D.; Wen, J.; Lloret, J. Busy tone-based channel access control for cooperative communication. *Trans. Emerg. Telecommun. Technol.* **2015**, *26*, 1173–1188. [CrossRef]

27. Chen, Y.; Liu, J.; Jiang, X. Throughput analysis in mobile ad hoc networks with directional antennas. *Ad Hoc Netw.* **2013**, *11*, 1122–1135. [CrossRef]

28. Besse, F.; Garcia, F.; Pirovano, A. Wireless Ad Hoc Networks Access for Aeronautical Communications. In Proceedings of the 28th AIAA International Communications Satellite Systems Conference, Anaheim, CA, USA, 30 August–2 September 2010; pp. 1–15.

29. Bianchi, G. Performance analysis of the IEEE 802.11 distributed coordination function. *IEEE J. Sel. Areas Commun.* **2000**, *18*, 535–547. [CrossRef]

30. He, X.; Kumar, R.; Mu, L.; Gjøsæter, T.; Li, F. Formal verification of a Cooperative Automatic Repeat reQuest MAC protocol. *Comput. Stand. Interfaces* **2012**, *34*, 343–354. [CrossRef]

Journal of
*Sensor and
Actuator Networks*

MDPI

Article

On-Line RSSI-Range Model Learning for Target Localization and Tracking

José Ramiro Martínez-de Dios [1,*], Anibal Ollero [1], Francisco José Fernández [2] and Carolina Regoli [3]

[1]   Robotics Vision and Control Group, Universidad de Sevilla, Escuela Superior de Ingenieros, c/Camino de los Descubrimientos s/n, 41092 Seville, Spain; aollero@us.es
[2]   Universidad de Sevilla, Escuela Superior de Ingenieros, c/Camino de los Descubrimientos s/n, 41092 Seville, Spain; fjfj@us.es
[3]   Central University of Venezuela, Ciudad Universitaria s/n, Caracas 1050, Venezuela; carolina.regoli@gmail.com
*   Correspondence: jdedios@us.es; Tel.: +34-954-487-357; Fax: +34-954-487-340

Received: 8 May 2017; Accepted: 5 August 2017; Published: 10 August 2017

Abstract: The interactions of Received Signal Strength Indicator (RSSI) with the environment are very difficult to be modeled, inducing significant errors in RSSI-range models and highly disturbing target localization and tracking methods. Some techniques adopt a training-based approach in which they off-line learn the RSSI-range characteristics of the environment in a prior training phase. However, the training phase is a time-consuming process and must be repeated in case of changes in the environment, constraining flexibility and adaptability. This paper presents schemes in which each anchor node on-line learns its RSSI-range models adapted to the particularities of its environment and then uses its trained model for target localization and tracking. Two methods are presented. The first uses the information of the location of anchor nodes to dynamically adapt the RSSI-range model. In the second one, each anchor node uses estimates of the target location—anchor nodes are assumed equipped with cameras—to on-line adapt its RSSI-range model. The paper presents both methods, describes their operation integrated in localization and tracking schemes and experimentally evaluates their performance in the *UBILOC testbed*.

Keywords: Wireless Sensor Network; RSSI; localization

1. Introduction

The Received Signal Strength Indicator (RSSI) is a main metric to estimate Quality of Service (QoS) of wireless links [1]. RSSI can be measured by the radio modules of most Commercial Off The Shelf (COTS) Wireless Sensor Network nodes with negligible cost in energy, delay and computational effort. The strength of the radio signals attenuates with distance as they propagate. Hence, a receiver can measure the RSSI of an incoming packet, estimate the distance and use it to find out the location of the emitter. Many RSSI-based localization methods have been researched and proposed over the years. One of the main drawbacks of using RSSI to estimate WSN link QoS in general—and also to estimate the distance between two WSN nodes—is that the strength of radio signals is highly affected by reflections and other interactions of the radio signal with the environment. These interactions are very difficult to be modeled and highly disturb the accuracy of RSSI-based localization techniques. In fact, many methods, e.g., *range-free* localization methods, compute location estimates without transforming RSSI into distance and hence, avoiding the inaccuracies of RSSI-range models.

Some techniques adopt a training-based approach in which they off-line learn the RSSI-range characteristics of the environment in a prior training phase and then, use the learned characteristics during the localization phase. In these methods the training phase is a long off-line time-consuming process. Besides, the training phase must be repeated if there have been changes in the environment.

This paper presents methods that address the learning of RSSI-range characteristics of the environment as an on-line process, i.e., training is performed during localization, enabling high adaptability to scenario changes. In this paper two methods are presented and compared. The first one uses the information of the location of the static nodes to dynamically adapt the RSSI-range model. The second one uses information gathered by nodes equipped with additional sensors—cameras—to continuously learn accurate RSSI-range models adapted to the local environment of the emitter and receiver nodes. The presented methods are efficient, scalable, operate in real-time and only require a few measurements to compute updated RSSI-range models. They are integrated in a target localization and tracking scheme that employs the trained RSSI-range models, significantly reducing target tracking errors. The paper presents both on-line RSSI-range model learning methods, describes their operation and experimentally evaluates their performance in the *UBILOC Testbed*.

This paper is structured as follows. Section 2 summarizes the related work. The problem formulation and general description of the presented methods are presented in Section 3. The two on-line RSSI-range model learning methods are described in Sections 4 and 5. Section 6 presents some experiments and discusses on the presented methods and results. The conclusions are in Section 7.

2. Related Work

RSSI is the most widely researched and employed measurement in localization and tracking in WSN. A high number of RSSI-based localization systems have been developed. *Range-based* methods, such as multilateration [2] or least squares [3], among many others, use RSSI measurements to estimate the distance to anchor nodes. Reflections and other interactions with the environment make RSSI-range models very dependent on the specific setting, making them unpredictable. Most range-based methods adopt a general RSSI-range model—such as references [4,5]—that assume free-space radio propagation and consider these perturbations as noise. Range-free methods, such as ROC-RSSI [6] or APIT [7], use geometrical considerations for localization instead of RSSI-range models. Although more robust to perturbations in the radio channel, their localization errors are usually higher than those of range-based techniques.

Another approach is to learn RSSI-range characteristics from the environment. In general, RSSI training-based localization and tracking includes two phases: (i) the training phase, where the RSSI-range characteristics of the environment are captured using measurements gathered in the scenario and represented typically in maps and; (ii) the localization phase, where the location estimation is performed using the RSSI map and the current RSSI measurements. The training phase is an off-line process that needs to be repeated if there have been changes in the wireless propagation environment. RSSI training based techniques can be broadly divided into deterministic and probabilistic methods.

Deterministic RSSI training-based localization techniques include references [8,9]. In the training phase the environment is subdivided into small cells and, RSSI measurements are gathered in these cells from several anchor nodes, which location is assumed known. In the localization phase the RSSI measurements from the target are gathered and the cell in the RSSI map that best fits with the current measurements is selected as the target location. Probabilistic RSSI training-based localization methods include references [10,11]. A probability distribution of the target location is defined over the area of the environment. The goal of the localization phase is to reach a single mode for this distribution, which is the most likely location of the target. In these methods a Bayesian belief network is established and trained with a preset number of discretized location possibilities (cells).

The aforementioned RSSI training-based localization methods rely on long training phases where a RSSI map of the entire target area is built with some spatial precision. This precision will have a great impact on the accuracy of the localization estimate. Such data collection/measurement for RSSI map building requires significant labor. Besides, the map must be recomputed whenever the wireless propagation properties may change. Thus, these methods cannot be applied in schemes that require quick deployment or in dynamic environments with frequent changes.

Modeling radio propagation characteristics with RSSI maps poses memory footprint issues when implementing these methods in COTS nodes. Some methods aim at building RSSI-range models trained for an specific environment. In reference[12] the authors evaluate ray-tracing techniques that are used to obtain indoor propagation models. In reference [13] a statistical approach is employed to build a wireless map based on statistical properties of the area. An even simpler and more practical approach is adopted in references [14,15]. In these methods RSSI-range models are computed using regressions but model fitting is designed as an off-line process, lacking flexibility and adaptability to environment changes.

3. On-Line RSSI-Range Training Problem Formulation

The objective of the presented method is to on-line obtain RSSI-range models adapted to the specific scenarios and environments. The local environment around two nodes critically impacts on the RSSI measurements between them. In fact, if we take RSSI measurements between two nodes without changing their local environments, we can notice that all the RSSI values are very similar. As an illustration Figure 1 shows 40,000 different RSSI measurements between one node and three other anchor nodes taken during 16 h in the *UBILOC Testbed*. Despite the high number of measurements, the difference between them was lower than 3 dBm and their standard deviation was lower than 1.4 dBm, which evidences the low level of noise in RSSI measurements when the environment surrounding the emitter and the receiver is static. This experiment was repeated in several occasions at different days and hours obtaining similar results. Thus, we can conclude that the main cause in the variability of RSSI measurements is not the random measurement noise itself, but changes in the environment surrounding the emitter and the receiver. Below a justification based on radio propagation is summarized.

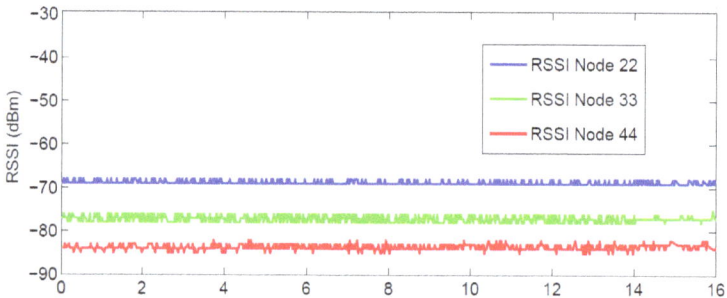

Figure 1. Result of RSSI variability experiments between three different anchor WSN nodes and the same receiver.

Most localization methods use RSSI-range models that assume free-space propagation [4,5], and adopt expressions similar to the following, see Appendix A:

$$RSSI(D) = a \log D + b, \tag{1}$$

where $RSSI(D)$ is the RSSI measured by a node from the packets it receives from the emitter node located at distance D from the receiver. a and b are the model parameters. This model assumes that the receiver sees only the direct radio signal transmitted by the emitter and, in this case, the RSSI measurement noise is statistically independent, as in the case of many other measurement processes. It should be noticed that the objective of model in (1) is to numerically relate RSSI and range. It can be easily derived from a general free-space radio propagation model.

However, the above model is ideal. In real cases the presence of reflectors in the environment creates multiple paths that a transmitted signal can traverse. As a result, the receiver sees the superposition of multiple copies of the transmitted signal, each of them traversing a different path. Each signal copy will experience different attenuation, delay and phase shift while traveling from the emitter to the receiver. The signal received is the composition of the direct signal and the contribution of these multiple copies. Hence, the signal strength of the resulting composition in a given environment is extraordinarily difficult to be modeled.

The most widely used approach in target localization and tracking methods is to adopt the model in (1)—i.e., propagation assuming no multi-path—and to consider the effects of multi-path propagation as noise. However, this noise is not originated by the randomness of the measurement process but by the influence of the particular environment surrounding the emitter and the receiver. This noise involves large errors in RSSI-range models and originates large localization and tracking errors. In recent years some few techniques for target localization under realistic correlated RSSI noise have been proposed [16] but they usually have bad scalability and involve computational loads that can be hardly implemented in COTS WSN nodes.

This work proposes methods that aim at capturing the overall effects of the full radio propagation using simple RSSI-range models that are trained continuously (on-line) during target tracking using updated RSSI measurements gathered by the anchor nodes deployed at the environment. In the presented methods each node dynamically trains its RSSI-range model using only the M most recent RSSI measurements. $M = 7$ was taken in the experiments. During this short interval, assuming moderate target motion, the environment—and hence the radio propagation conditions—can be considered static or almost static.

The paper presents two methods for on-line RSSI-range modeling and compares their performance in a target localization and tracking scheme, see Figure 2a. In the first method, described in Section 4, each anchor node uses inter-anchor RSSI measurements to dynamically train the RSSI-range model. In the second one, described in Section 5, each anchor node uses information gathered by external sensors—cameras—to continuously update and train RSSI-range models adapted to the environment conditions. In both cases, once computed, the trained RSSI-range model is used as the observation model in the target localization and tracking scheme in Figure 2a. Figure 2b shows an example of the geometrical setting of the experimentation scenarios. The optical axes of the cameras of the sensor nodes in the method described in Section 5 are shown.

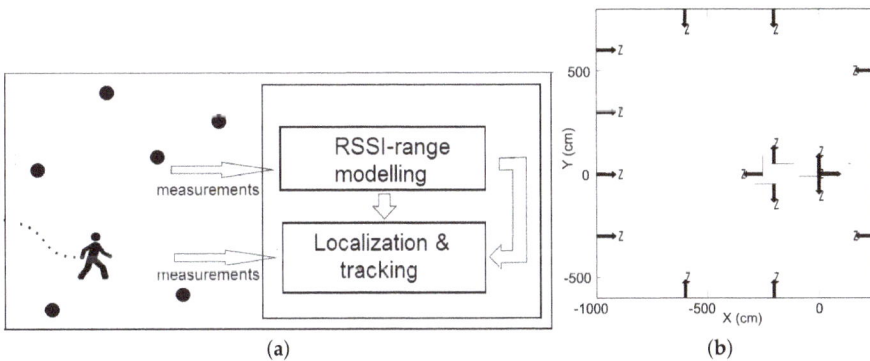

Figure 2. (**a**) General target localization and tracking scheme with on-line trained RSSI-range models; (**b**) Geometrical setting of the scenario.

4. On-Line Training Using Inter-Anchor RSSI Measurements

In this method, each anchor node dynamically computes its RSSI-range model using the RSSI measurements from the packets it interchanges with other anchor nodes. These measurements are used to on-line perform the RSSI-range model regression. The method is general and only assumes that each node knows its own location, which is reasonable and does not constrain its generality.

4.1. Inter-Anchor RSSI-Range Model Training

This method adopts the widely assumed model expressed in (1) in which RSSI-range is taken as a linear function between the RSSI and the logarithm of the distance D. Assuming a set of measurements of the type $\{(RSSI_k, \log D_k)\}$, the model in (1) can be fitted by performing a simple linear least squares regression:

$$a = \frac{\sum_k RSSI_k \log D_k - \overline{RSSI} \sum_k \log D_k}{\sum_k (\log D_k)^2 - \overline{\log D} \sum_k \log D_k},$$ (2)

$$b = \overline{RSSI} - a(\overline{\log D}),$$ (3)

where $\overline{RSSI}$ and $\overline{\log D}$ stand for the mean of $RSSI_k$ and of $\log D_k$, respectively.

A default RSSI-range model, with predefined values for the parameters a and b is set for every node before starting the tracking application. Figure 3 shows the RSSI-range model computed by fitting (1) with RSSI measurements from every pair of anchor nodes deployed in the *UBILOC Testbed*. It is taken as the default RSSI-range model in the performed experiments.

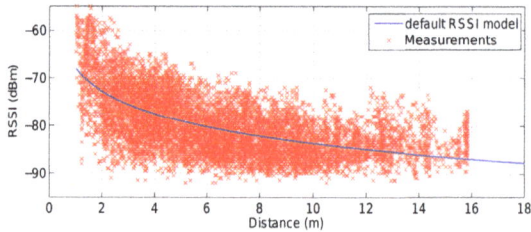

Figure 3. RSSI-range model experimentally obtained in the *UBILOC Testbed*.

The operation of the RSSI-range training stage is as follows, see Figure 4. Each anchor node i starts periodically broadcasting *RegrRequest* packets. If anchor node j receives a *RegrRequest* packet, it responds transmitting a *RegrResp* packet containing its ID and its location L_j. Node i receives the *RegrResp* packet, measures its $RSSI_{ij}$, extracts L_j and computes $D_{ij} = \|L_i - L_j\|$. Hence, anchor node i obtains a measurement pair $(RSSI_{ij}, \log D_{ij})$ from each packet it receives from node j. If node i has sufficient measurements it can fit its RSSI-range model $RSSI_i(D)$ using (2) and (3).

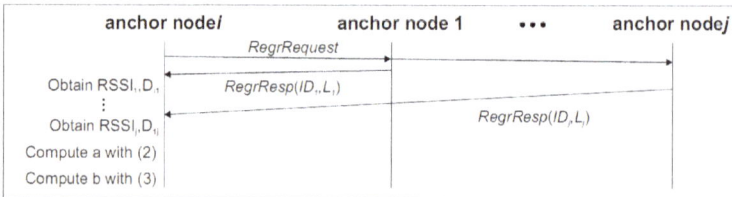

Figure 4. Operations and communications between anchor node i and the rest of anchor nodes for inter-anchor RSSI-range model training.

Only the *M* more recent measurements are used for RSSI-range training. As anchor node *i* gathers more measurements, it periodically recalculates its RSSI-range model. This has two positive effects. First, the model dynamically adapts to changing environment conditions. Second, it decreases the sizes of the measurement set used in the regression, reducing the computational burden.

4.2. Outlier Rejection

Least Squares regression can cope with Gaussian noise but performs badly with non-Gaussian noise measurements. Some authors like references [17,18] reported that RSSI noise is not always Gaussian. The presented method uses the RANSAC (RANdom SAmple Consensus) algorithm [19] to filter out highly noisy measurements, —outliers.

RANSAC is an efficient algorithm to estimate parameters of a mathematical model from a set of measurements that can contain outliers. It requires a set of measurements $\{(RSSI_k, \log D_k)\}$ and a parameterized model, such as (1). RANSAC iterates selecting a random subset of the measurements as hypothetical inliers. At each iteration the model is fitted with the hypothetical inliers, and the rest of the measurements are tested against the fitted model. If an observation fits well to the estimated model, it is considered to be a hypothetical inlier. If a sufficient number of measurements have been classified as hypothetical inliers, the estimated model is considered good and the model is re-estimated using all hypothetical inliers. Finally, the model is evaluated by estimating the error of the inliers relative to the model. This procedure is repeated. The model with the lowest error after the iterations is selected.

Figure 5 shows the model obtained with and without using RANSAC in an experiment performed in the *UBILOC testbed*. The figure shows 200 RSSI measurements taken by one static node from 5 anchor nodes. Although the RSSI measurements from each node showed low variability, the model fitting error can be high when considering all the measurements. The model obtained without RANSAC using all measurements was $a = -9.51$ and $b = -5.34$, see Figure 5 in dashed line. The mean fitting error was 2.23 dBm. The outliers detected with the RANSAC algorithm are represented with asterisks. The model obtained with RANSAC using only inliers measurements—shown in solid line—yielded $a = -10.302$ and $b = -1.678$, with a mean fitting error of 0.49 dBm. The differences between both models can be noticed in Figure 5 .

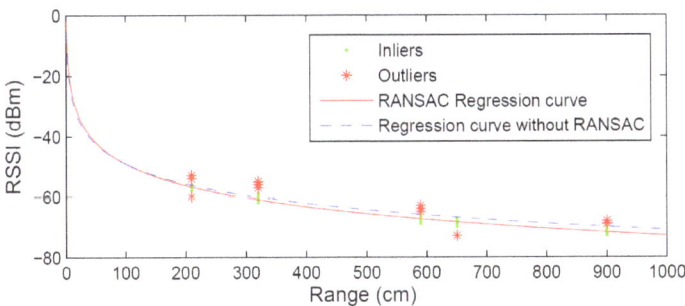

Figure 5. Experimental RSSI-range models obtained with and without RANSAC.

This method is computationally efficient and can be implemented by COTS WSN nodes. Each anchor node needs measurements from only two anchor nodes to on-line train its RSSI-range model. Of course, the model trained with few measurements will have lower accuracy that those trained with higher number of measurements. This method trains the RSSI-range model using measurements interchanged between anchor nodes. Hence, the trained model will capture the RSSI-range behavior at few discrete locations in the environment but, will not be able to capture the influence of the local surroundings of the moving target. This is the main advantage of the on-line training using target-anchor RSSI measurements, which is described in the next section.

5. On-Line Training Using Target-Anchor RSSI Measurements

In this method, anchor nodes are equipped with additional sensors that enable obtaining updated estimates of the target location. Each anchor node knows its own location and uses the estimation of the current target location to continuously adapt its own RSSI-range model. The resulting trained models are adapted to the local environments of the emitter (target) and receiver (anchor node). In this method each anchor node is assumed equipped with a camera and endowed with image processing and perception functions capable of estimating the current target location. Cameras have high energy consumption. Hence, in our method the cameras are only used to train the RSSI-range model and, not to perform target localization and tracking as e.g., reference [20], which would involve high energy consumption. The method includes mechanisms in which cameras are turned on to analyze the accuracy of the current trained RSSI-range model and update it if necessary. After that, the cameras are turned off to save energy.

5.1. Target-Anchor RSSI-Range Model Fitting

Let $RSSI_{i,k}$ be the RSSI measured by anchor node i from the packets it receives from the target. First we assume that each anchor node knows the updated location of the target. This will be relaxed in the next subsection. Assuming that each anchor node knows its location, anchor node i can collect a set of measurements $(RSSI_{i,k}, d_{i,k})$, where $d_{i,k}$ is the distance between the node i and the current target location. The objective is to dynamically obtain a RSSI-range model for anchor node i adapted to the local environment of the target at time k. This RSSI-range model can be approximated as linear: the anchor node i trains a RSSI-range model that is valid only in the proximity of the target current location. Locally the logarithm can be well approximated by a linear model, which reduces burden and simplifies its statistical-based analysis and filtering. In order to cope with the target motion, models are trained only with the M most recent pairs collected. Low values of M are typically selected. If the target moves, high M can involve RSSI measurements with different target local environments.

Hence, in this method a simple linear RSSI-range model for node i ($RSSI_{i,k} = a_i d_{i,k} + b_i$) (see Appendix A) is efficiently fit using Least Squares with measurement pairs ($RSSI_{i,k}, d_{i,k}$):

$$a_i = \frac{\sum_k RSSI_{i,k}d_{i,k} - \overline{RSSI_i}\sum_k d_{i,k}}{\sum_k (d_{i,k})^2 - \overline{d_i}\sum_k d_{i,k}}, \tag{4}$$

$$b_i = \overline{RSSI_i} - (a_i\overline{d_i}), \tag{5}$$

where $\overline{RSSI_i}$ and $\overline{d_i}$ stand for the mean of $RSSI_{i,k}$ and of $d_{i,k}$.

The operation of the RSSI-range training is depicted in the diagram shown in Figure 6. The target starts periodically broadcasting *RegrRequest* packets. If anchor node i receives a *RegrRequest* packet, it measures its $RSSI_i$ and computes the range to the target $d_{i,k}$. Hence, anchor node i obtains a measurement pair ($RSSI_i, d_{i,k}$) from each packet it receives from the target. If anchor node i has sufficient measurements, it can fit its RSSI-range model using Eqautions (4) and (5).

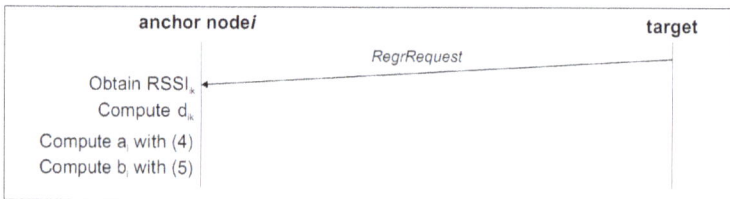

Figure 6. Operations and communications between anchor node i and the target in target-anchor RSSI-range model training.

Only the M more recent measurements are used for model training. As anchor node i gathers more measurements it recalculates a_i and b_i enabling dynamic adaptation to the changing environment and reducing the size of the training pair set.

5.2. RSSI-Range Model Update

Section 5.1 assumed that the actual location of the target is known by all anchor nodes. Of course, this is not the case and nodes should estimate the target location. This technique uses estimates of the target location obtained from the cameras adopting a simple *Maximum Likelihood* (ML) approach that computes the location of the target that best matches with the observations gathered by the different cameras. The method used is summarized below.

Each sensor node gathers one image and applies image processing techniques to segment the target in the image. Let $p_{i,k}^{im}$ be the location (in pixels) of the center of the region where the target at time k has been segmented in the image from camera i. Assuming the camera pin-hole model is known, from $p_{i,k}^{im}$ it is easy to compute $p_{i,k}^c$, the location of the target on the scenario measured by camera i expressed in the reference frame local of camera i. Using T_i, the transformation matrix from the reference frame of camera i w.r.t. a common reference frame, $p_{i,k}^c$ can be transformed to $p_{i,k}^s$, which represents the location of the target measured by camera i in a common reference frame of the scenario. It is assumed that $p_{i,k}^s$ contains errors that can be modeled as Gaussian statistically independent noise with zero mean noise and covariance matrix $Cov_{i,k}$. Thus, the ML technique estimates the location of the center of the target as a Gaussian distribution with mean P_k and covariance Cov_k as follows [21]:

$$P_k = (\sum_{i=1}^{N} Cov_i^{-1}) \sum_{i=1}^{N} (p_{i,k}^s Cov_{i,k}^{-1}),\tag{6}$$

$$Cov_k = 1/(\sum_{i=1}^{N} Cov_{i,k}^{-1})\tag{7}$$

Each sensor node i broadcasts its values $p_{i,k}^s$ and $Cov_{i,k}$ and combines them with those received from other nodes using Equations (6) and (7). Figure 7 shows an example of the ML technique with two cameras. The distributions of $p_{i,k}^s$ for Camera1 and Camera2 are in blue color. For both cases the values of $Cov_{i,k}$ along the camera axes were assigned high to reflect the high uncertainty in the depth reconstruction when computing $p_{i,k}^c$. The resulting probabilities of the target location distribution (P_k), in red color, are significantly higher, which denotes an increment in location accuracy. For more details on the method, please refer to reference [21].

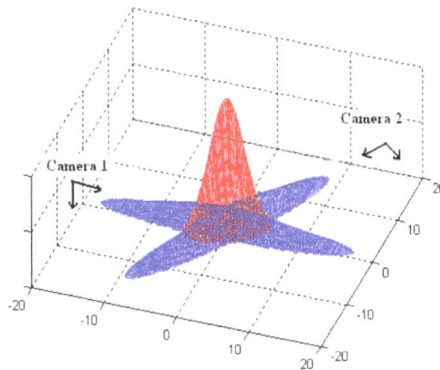

Figure 7. Simple example of the ML technique.

The camera measurements are used only for RSSI-range model training. Each anchor node that is tracking a target periodically performs RSSI-range model update. During RSSI-range model update each node analyzes the accuracy of its RSSI-range model and updates it if its accuracy is found unsuitable. The operation is as follows. The target periodically transmits a *RSSIRangeModelUpdate* packet. In the experiments this frequency was 0.33 Hz. When an anchor node *i*—located at L_i—receives it, it turns on its camera, gathers images and estimates the target location P_k. The error of its RSSI-range model can be computed comparing the distance to the target estimated using camera measurements and the distance predicted by its RSSI-range model $d_{i,k}$:

$$e_{i,k} = \frac{(RSSI_{i,k} - b_i)}{a_i} - \|P_k - L_i\| \tag{8}$$

The RMS error in the most recent N measurements is:

$$E_i = \frac{1}{N} \sum_{k=1}^{M} e_{i,k}^2 \tag{9}$$

E_i can be used to estimate the validity of the RSSI-range model. If $E_i < TH$, the RSSI-range model of anchor node *i* is considered valid: it is not updated and the cameras are turned off. Otherwise, the RSSI-range model is considered invalid and should be computed using new $(RSSI_{i,k}, d_{i,k})$ pairs with Equations (4) and (5). After model training, the cameras are turned off. *TH* is the bound in the admissible error in the RSSI-range model and depends on the accuracy requirements of the application. Low values of *TH* originate frequent RSSI-range model update, which improves model accuracy but involves gathering more camera measurements and hence, higher energy consumption. High values of *TH* involve lower frequency in RSSI-range model update and hence, lower energy consumption. In preliminary experiments $TH = 0.35$ m^2 was found a good trade-off between model accuracy and energy consumption and it was adopted in the experiments shown in this paper.

Figure 8 shows two trained RSSI-range models for anchor node *i* computed by the presented method when the target is at two different locations along the target path in one experiment performed in the *UBILOC Testbed*. The improvements in accuracy and fitting error over the default RSSI-range model in Figure 3 are evident. In this method the RSSI-range models are specific for each anchor node and trained to be accurate for the local surroundings of the target and, hence they are significantly more accurate than the default model.

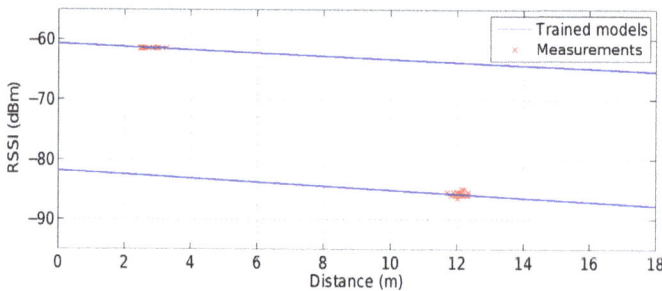

Figure 8. Trained RSSI-range linear models obtained in an experiment in the *UBILOC Testbed* using target-anchor RSSI measurements.

It can be noticed in Figure 8 that, as expected, very few outliers were found when fitting the RSSI-range model. This result is consistent: the RSSI-range model is fitted to the local surroundings of the emitter and receiver, which can be considered static for all the M measurements used in the $(RSSI_{i,k}, d_{i,k})$ pairs.

It should be noticed that the inaccuracies in the estimation of P_k can affect as perturbations in the RSSI-range model training. The proposed method includes a mechanism to specifically consider these errors. It is a statistical-based supervisor that estimates the variance of the error of the trained RSSI-range model. It operates as follows. Assume that $d'_{i,k} = d_{i,k} + u_i$, where $d'_{i,k}$ is the distance from anchor node i to the target location P_k estimated with the cameras, $d_{i,k}$ is the actual distance and u_i is the error in the estimation of $d'_{i,k}$, which is highly related to the error in the estimation of the target position using camera measurements. Assume that $RSSI_{i,k} = a_i d_{i,k} + b_i$ is the exact RSSI-range model and that the measured RSSI is $RSSI'_{i,k} = RSSI_{i,k} + v_i$, where v_i is the RSSI measurement error. The training method uses pairs $\{(RSSI'_{i,k}, d'_{i,k})\}$ to fit the model $RSSI'_{i,k} = a_i d'_{i,k} + b_i + v_i$. It is easy to check that the following expression holds:

$$RSSI'_{i,k} = a_i(d_{i,k} + u_i) + b_i + v_i = RSSI_{i,k} + a_i u_i + v_i \tag{10}$$

In virtue of the Central Limite Theorem, u_i and v_i are assumed Gaussian White noises with zero means and variances $\sigma^2_{u_i}$ and $\sigma^2_{v_i}$. Hence, the variance of the error of the trained RSSI-range model is:

$$\sigma^2_{tm,i} = a_i^2 \sigma^2_{u_i} + \sigma^2_{v_i} \tag{11}$$

Hence, it is very efficient to compute $\sigma^2_{tm,i}$. It depends on: $\sigma^2_{v_i}$; a_i, the slope of its RSSI-range model; and $\sigma^2_{u_i}$, the variance of the error in the estimation of $d'_{i,k}$, which can be easily computed from Cov_k, the covariance of the error in the target location estimated with camera measurements. Trained RSSI-range models with higher a_i are more sensitive to target location errors. Cases where $1/a_i = 0$ are approximated by $1/a_i = \epsilon$ and b_i is recomputed using Equation (5) with the new a_i.

The proposed method includes an efficient mechanism to evaluate the effect of estimation errors in the trained RSSI-range model. If $\sigma^2_{tm,i}$ is higher than the variance of the default RSSI-range model, it means that the trained model is too noisy and it is considered invalid. In these cases, the default RSSI-model is preferred instead. As a result, badly trained RSSI-range models can be easily detected and, in those cases the default RSSI-range model is employed instead.

The presented technique allows obtaining RSSI-range models with accuracies similar to cameras but with very low energy consumption. Besides, it is endowed with mechanisms to self-detect and correct model training errors.

6. Experimental Section

In this section, the presented on-line RSSI-range modeling techniques are evaluated in real experiments. Both methods were inserted in the target localization and tracking scheme depicted in Figure 2 and experimented in the *UBILOC Testbed*. The *Localization & tracking* module in Figure 2, which estimated the target position using RSSI measurements, was implemented by a Kalman Filter (KF). A standard Kalman Filter was implemented. The only difference was that the observation model employed was not constant: it used the RSSI-range model trained on-line (during target tracking) by the two presented methods. Kalman Filters have been widely-employed in RSSI-based localization and tracking. For brevity, they are not presented in this paper.

6.1. The UBILOC Testbed

The Ubiquitous Localization Testbed (*UBILOC*) is a tool designed for experimenting localization and tracking techniques in ubiquitous computing systems, see Figure 9a. It emulates a typical office smart environment scenario and is deployed in a room of more than 500 m^2 in the basement of the School of Engineering of Seville. *UBILOC* was developed on top of the *CONET Integrated Testbed* [22,23], and inherits from it its main characteristics.

UBILOC includes sensors and systems widely used in localization and tracking in ubiquitous computing systems. It includes a WSN with static and mobile nodes. The mobile nodes are used as tags to be mounted on the targets to be localized and tracked. The static WSN nodes are deployed at 21 predefined locations hanging at 1.7 m height from the floor. It also includes a Wireless Camera Network (WCN). Each node of the WCN is composed by a *CMUcam3* camera module [24] connected to a *TelosB* WSN node, see Figure 9b. *CMUcam3* camera modules are endowed with embedded programmable image processing capabilities. Each *CMUcam3* module captures RGB images and applies simple image processing methods. The results of the local image processing methods executed at each camera are transmitted to the *TelosB* node using a simple bidirectional protocol.

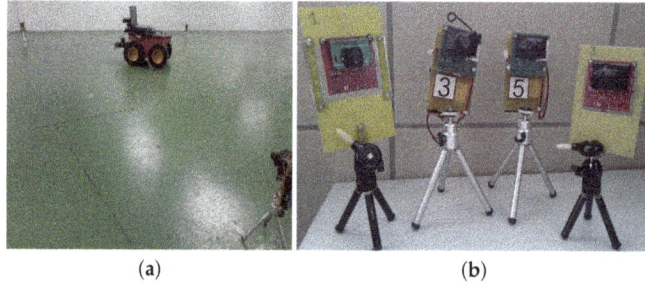

(a)   (b)

Figure 9. (**a**) Picture of the *UBILOC Testbed* during an experiment; (**b**) Nodes of the WCN: each node is composed of a *CMUcam3* module and a *TelosB* WSN node.

UBILOC also includes a set of mobile robots that are used as targets to be localized and tracked in the experiments. They enable high repeatability in localization and tracking experiments. Each robot is equipped with one 2D laser range finder and one Microsoft Kinect and an IEEE 802.11 Wireless bridge. Each robot is capable of accurately computing its own location and orientation using the Adaptive Monte Carlo Localization algorithm (AMCL) [25]. These location estimations are taken as the ground truth for evaluating target localization and tracking techniques in indoor experiments.

6.2. Proof of Concept

This section illustrates with a simple example the operation of the on-line target-anchor RSSI-range model training. The operation of the inter-anchor RSSI-range model technique is rather simple and it is not presented.

Figure 10 shows the result of a target localization and tracking experiment performed with the scheme in Figure 2 using the target-anchor RSSI-range model training method. The ground truth robot location is represented in blue and the estimated target location is in red. The camera local frames are also shown. Figure 10 shows in green the path followed by the robot between $t = 550$ s and $t = 570$ s. At $t = 550$ s anchor nodes ID1, ID2 and ID3 had their cameras off. This time is taken as $k = 0$ in Figure 11, which shows in a simplified way the operation of the target-anchor RSSI-range model training between $k = 0$ and $k = 20$. Figure 11a shows whether each of the involved anchor node is at its RSSI-range model update stage—high value—or not—low value. Figure 11b shows E_i, the RMS error of the target-anchor RSSI-range model of each node along time. TH is also shown in Figure 11b.

Nodes ID1, ID2 and ID3 updated their RSSI-range model at $t = 540$ s. At time $t = 574$ s—i.e., $k = 4$—they received a *RSSI Range Model Update* packet from the target and started their RSSI-range model update stage. Each node turned on its camera, gathered visual measurements, used these measurements to estimate the target location and, with them computed E_i, the RMS error of its RSSI-range model. If $E_i < TH$ the model for node i is considered valid: it finishes its RSSI-range model update stage and turns its camera off. This is the case of ID1, $E_i = 0.31$ as shown in Figure 11b. ID1 finishes its RSSI-range model update stage and turns off its camera. Between $k = 0$ and $k = 8$, the cameras of nodes ID1, ID2 and ID3 are kept inactive. They cannot compute E_i and it is not represented in Figure 11b.

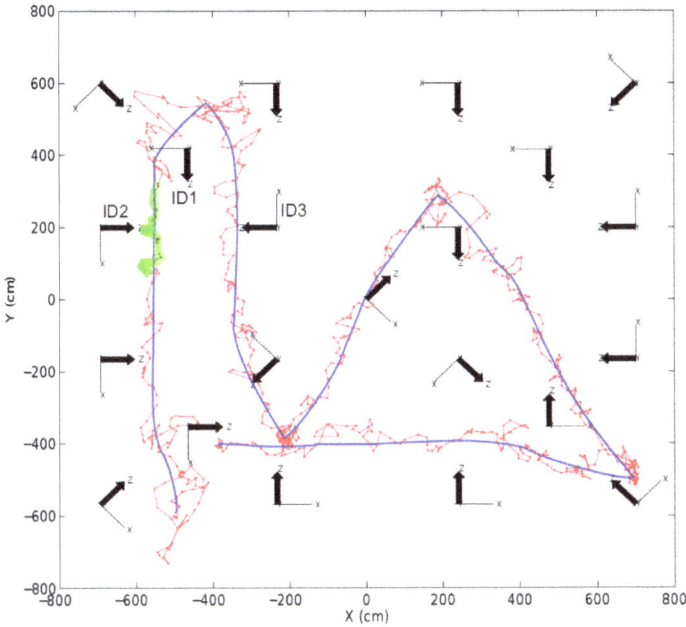

Figure 10. Result of an experiment performed with the scheme in Figure 2 using target-anchor RSSI-range model training: ground truth in solid (**blue**) and estimated location (**in red**).

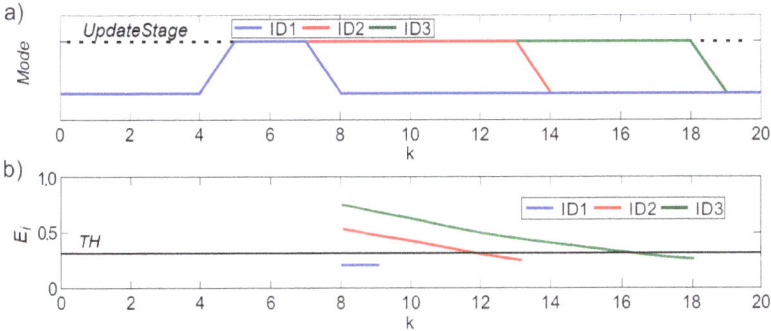

Figure 11. Performance of nodes ID1, ID2 and ID3 in the experiment shown in Figure 9 between second $t = 550$ s ($k = 0$) and second $t = 570$ s ($k = 20$): (**a**) mode of anchor nodes ID1, ID2 and ID3; (**b**) values of E_i of each node along time.

If $E_i > TH$ the model for node i is considered invalid: it should update its RSSI-range model. This is the case of anchor nodes ID2 and ID3. With each new target location estimate, each of these nodes generated a new pair ($RSSI_{i,k}$, $d_{i,k}$) and used the M most recent pairs to train its RSSI-range model using Equations (4) and (5). Every time the RSSI-range model of node i is updated, the new E_i is computed to evaluate the accuracy of the new model. Nodes ID2 and ID3 kept gathering visual measurements and updating their RSSI-model until they satisfied $E_i < TH$. At that time the updated model was considered valid: the node finished its RSSI-range model update stage and turned its camera off. In the example in Figure 11 the RMS error of the RSSI-range model of ID3 was higher than that of ID2. ID2 needed to update the pair set $(RSSI_{i,k}, d_{i,k})$ with few measurements to compute an updated RSSI-range model that satisfied $E_i < TH$. On the other hand, E_i for node ID3 was higher and required more measurements. At $k = 17$ its trained RSSI-range model achieved $E_i < TH$, ID3 finished its RSSI-range model update stage and turned its camera off .

In the presented target-anchor RSSI-range model training technique the cameras are active only while gathering measurements that are used for the regression of the RSSI-range model, which represents a low percentage of the time in the experiment. During the rest of the time the cameras are kept inactive saving energy.

6.3. Evaluation and Comparison

Both presented on-line RSSI-range modeling methods were employed in the scheme depicted in Figure 2 and evaluated in target localization and tracking experiments. The method using inter-anchor measurements was implemented as described Section 4. From now on it will be called M1. The method using target-anchor measurements was implemented as described Section 5. It is called M2.

Consider the scenario shown in Figure 12, where 21 camera-nodes were deployed hanging from the ceiling (3 m height). The optical axis of all the cameras are also shown. The roll angle of all the cameras were zero. The target was a robot, which motion was controlled using the Player/Stage random walk functionality, in which it is given pseudo-random velocity commands. Figure 12 shows in black the trajectory actually described by the robot. The target location estimated using the tracking scheme in Figure 2 using M1 and M2 are represented respectively with dashed red lines and blue points. In this experiment M1 achieved an average localization error of 1.83 m and, M2, of 0.47 m.

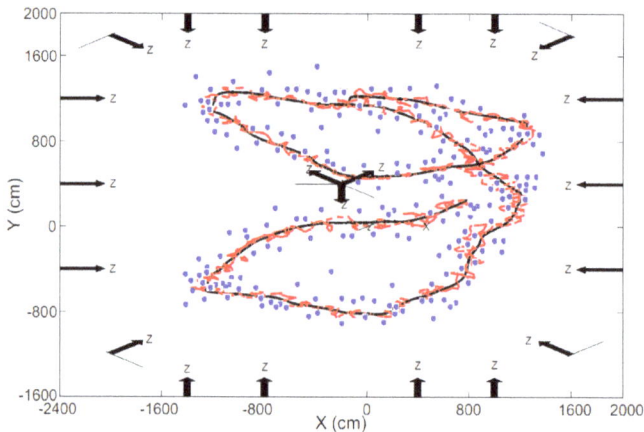

Figure 12. Result of a target localization and tracking experiment performed in *UBILOC*: ground truth is the solid black line, the target location estimated using M1 and M2 are represented respectively with dashed red lines and blue points.

The accuracy of the presented on-line RSSI-range model training techniques was evaluated in a set of experiments with 70 random settings and robot trajectories. The camera nodes were deployed on the floor randomly using a uniform distribution in the range of the room size. The camera roll and pith angles of all the cameras were zero and the yaw angles were selected such that the optical axes pointed at the center of the room. We performed the experiments, logged all RSSI measurements and off-line processed the logs with the scheme in Figure 2 using three different approaches: the default RSSI-range model, method M1 and method M2. Figure 13 compares the mean localization error obtained using the three approaches.

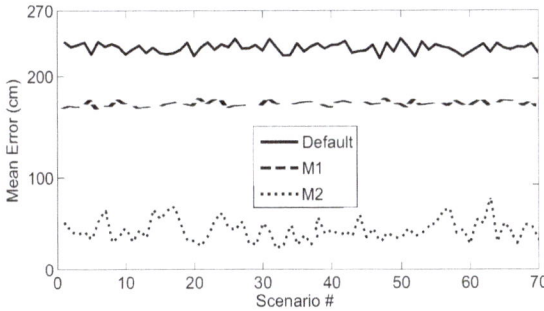

Figure 13. The mean localization error obtained by the scheme in Figure 2 when using the three approaches: Default, M1 and M2 in 70 different experiments performed in *UBILOC*.

In average with M1 the target localization and tracking scheme had a localization error 18% lower than when using the default RSSI-range model. Also, with M2 the localization error was 77% lower than that of the scheme that uses the default model. The target-anchor RSSI-range modeling provides very accurate models. The experiments were performed assuming realistic conditions. For instance, Figure 14 shows the Packet Reception Rate (PRR) measured experimentally during the experiments.

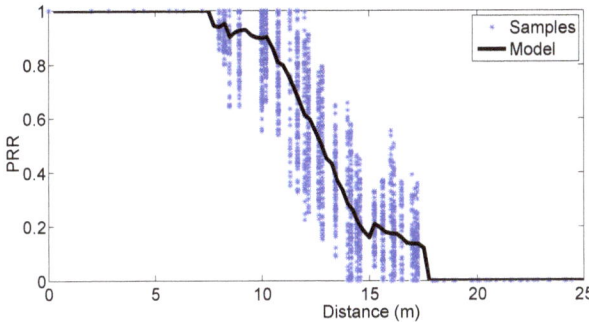

Figure 14. Packet Reception Rate obtained experimentally in the described experiments.

Figure 15 shows the cumulate target localization errors obtained in the three cases: default, M1 and M2. M2 had significantly higher accuracy: the mean error was 53 cm and the error was lower than 97 cm in 80% of the samples. Errors were significantly higher in M1 and also in the default model.

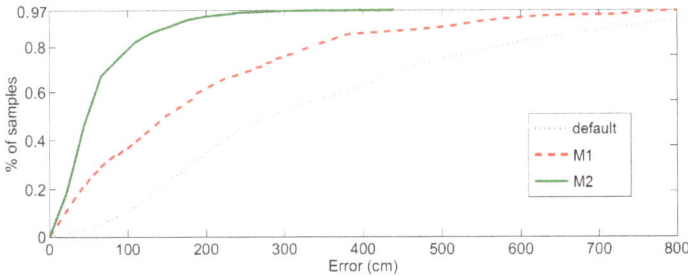

Figure 15. Cumulative target localization errors obtained by the scheme Figure 2 in the above experiments when using the default model, method M1 and method M2.

M2 provides accurate models because it estimates target location with cameras and trains the RSSI-range model dynamically considering the local surroundings of the target and of the anchor node. Besides, it includes self-corrective tools that improve robustness against camera errors. Table 1 summarizes the performance of the target localization and tracking scheme with the default model, the models resulting from methods M1 and M2. In M2 the cameras were active during an average of 11.5% of the total duration of the experiment. Considering the energy consumption of *TelosB* WSN nodes and *CMUcam3* modules, which can be found in their data-sheets, the energy consumed by M2 was 140.75 mW, 61% higher than in case of using the default model and method of M1, which did not use cameras. M2 obtains localization and tracking errors similar to vision-based methods but requires having the cameras active only during a small fraction of the experiment duration and hence, has significantly lower energy consumption than tracking systems based solely on cameras.

Table 1. Performance when using the default model, and those resulting from methods M1 and M2.

	Default	M1	M2
Average localization error (m)	2.32	1.89	0.53
Average time cameras are active (% of the experiment time)	0	0	11.5%
Average sensor node energy consumption (mW)	66	66	140.75

6.4. Discussion

M, the number of measurement pairs used in RSSI-range model regression, is critical both for M1 and for M2. On one hand, higher values of *M* reduce inaccuracies in model fitting. On the other hand, using older measurements reduces their capability to adapt to the changes in the scenario, particularly in M2, which uses target-anchor measurements and the target is usually in motion.

Figure 16 shows the average target tracking error in the above experiments when using M1 with different values of *M*. In M1 the models trained with fewer measurement pairs were less accurate than those fitted with higher *M*. With less than 3 measurements from each anchor node, the trained RSSI-range model obtained errors that are similar to those when using the default RSSI-range model. It was necessary to select at least $M = 5$ to exploit the benefits of on-line fitting. However, the target tracking error did not decrease when using $M > 7$. Taking higher *M* involved higher computational burden but did not increase the accuracy of the RSSI-range model.

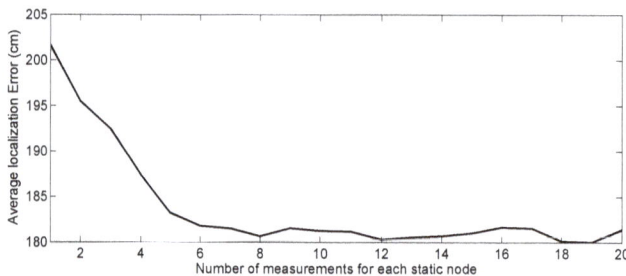

Figure 16. Average tracking error in Figure 2 with M1 when using different number of measurement pairs.

Figure 17 shows the average target tracking error when using M2 with different values of *M*. In M2 linear RSSI-range models enable accurate fitting even when using very low number of measurement pairs. With higher *M* the RSSI-range models were less accurate since the target was in motion and with a higher number of measurement pairs the model was not fitted to the surroundings of the target but to a wider area, loosing accuracy.

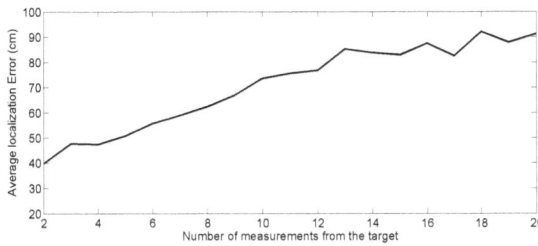

Figure 17. Average tracking error in Figure 2 with M2 when using different number of measurement pairs—*M*.

In the following, the complexity (number of operations) of M1 and M2 are estimated. Assume that in M1 the measurements of K anchor nodes are used in the training of the RSSI-range model of each anchor node. M is the number of measurement pairs employed in the regression. Also, the RANSAC algorithm performs I iterations and F is the fraction of samples taken for the RANSAC initial fitting. The complexity of M1 without outlier rejection is $O(2KM)$. The complexity of outlier rejection in M1 increases proportionally to the number of RANSAC iterations I. The complexity of M1 with RANSAC is $O(2IFKM)$. The complexity of scheme M2—without including the number of operations to compute P_k—is $O(M)$.

We take typical values for the above parameters in the experiments performed in *UBILOC*: $K = 5$, $M = 7$, $I = 4$ and $F = 0.2$ for M1 and, $M = 4$ for M2. Method M1 without RANSAC requires $NO_{M1} = 73$ operations and M1 with RANSAC requires $NO_{M1-R} = 537$ operations. The use of RANSAC for outlier rejection requires significant burden and the improvement in accuracy can be interesting in some applications. Method M2 requires only $NO_{M2} = 11$ operations. M2 is very efficient but requires accurate target location estimates obtained with other sensors such as cameras. M2 is very interesting if anchor nodes are equipped with cameras.

7. Conclusions

RSSI is the most widely used measurement for localization and tracking in ubiquitous computing systems. RSSI can be measured by most radio modules used in COTS WSN nodes without incurring in extra delay, computational or energy consumption costs. However, RSSI measurements are highly affected by perturbations originated by interactions of the radio signal with the environment, which significantly disturb the accuracy of RSSI-based localization techniques.

Most range-based localization methods use a general RSSI-range model that assumes free-space radio propagation and considers these perturbations as noise. However, this noise is correlated and does not satisfy the hypotheses assumed by these techniques, involving high localization errors. Some techniques adopt a training-based approach in which they off-line learn the RSSI-range characteristics of the environment in a prior training phase. However, the training phase is a long off-line time-consuming process, which must be repeated if there have been changes in the environment, lacking adaptability.

This paper deals with methods in which each anchor node on-line learns its RSSI-range model adapted to the particularities of its environment and uses its trained model for localization and tracking. In the presented methods the training stage is performed during the localization stage, enabling adaptability to scenario changes. Two techniques are presented. The first method (M1) uses the information of the location of the anchor nodes to dynamically adapt the RSSI-range model. The second methods (M2) uses the target location estimated by anchor nodes equipped with cameras to on-line adapt the RSSI-range model. The presented methods operate in real-time and only require a few measurements to compute updated RSSI-range models.

J. Sens. Actuator Netw. **2017**, *6*, 15

The presented techniques have been integrated in a target localization and tracking scheme based on Kalman Filter and evaluated in experiments performed in the *UBILOC Testbed*. In average with M1 had a localization error 18% lower than when using the default RSSI-range model. M1 does not require camera measurements and had the same energy consumption as the default case. The localization error in M2 was 77% lower than when using the default model. In M2 the cameras were active during 11.5% of the total duration of the experiment, involving an energy consumption 61% higher than in M1 or in the default case. Hence, M2 obtains localization and tracking errors similar to vision-based methods but requires having the cameras active only during a small fraction of time, involving high energy savings w.r.t. systems based solely on cameras.

Acknowledgments: This work was developed mainly in the context of EU Project MULTIDRONE (H2020-ICT-2016-2017/H2020-ICT-2016-1). Partial funding was obtained from EU Project AEROARMS Ref. H2020-ICT-2014-1-644271 and the AEROMAIN project funded by the Spanish R&D plan (DPI2014-59383-C2-1-R). José Ramiro Martínez-de Dios acknowledges EU Project AEROBI funded under contract H2020-ICT-2015-1-687384.

Author Contributions: José Ramiro Martínez-de Dios was the main designer of the scheme and algorithms that led to this paper, coordinated the experiments and elaborated the major part of the manuscript. Anibal Ollero provided suggestions and corrections during the preparation of the manuscript. Francisco José Fernández and Carolina Regoli participated in the implementation and evaluation of the methods.

Conflicts of Interest: The authors declare no conflict of interest.

Appendix A. Derivation of RSSI-Models

In the basic radio free-space propagation model the following expression holds:

$$RSSI(D) = P0(D0) - 10n_p \log(D/D0), \tag{A1}$$

where $RSSI(D)$ is the value of RSSI in dBm at distance D, $P0(D0)$ is the strength of the transmitter in dBm at a reference distance $D0$, n_p is the path loss exponent. Mathematically, Equation (A1) can be expressed as:

$$RSSI(D) = -10n_p \log(D) + P0(D0) + 10n_p \log(D0) \tag{A2}$$

n_p depends on the environment. In a model training approach $D0$ and $P0(D0)$ are treated as unknowns that are fitted during training. In a model training problem, Equation (A2) can be expressed as:

$$RSSI(D) = a \log(D) + b, \tag{A3}$$

where $a = -10n_p$ and $b = P0(D0) + 10n_p \log(D0)$.

This RSSI-range model reflects the numerical relations between RSSI and range. Although it is simple for efficient RSSI-range model training, it is actually derived as a simplification of physical radio free-space propagation model as explained below. The linear RSSI-range model used in Section 5 is also an experimental RSSI-range model which objective is to numerically relate the values of RSSI and range. This RSSI-range model is computed continuously and hence it is enough to compute a simple model valid for the surroundings of the target.

References

1. Baccour, N.; Koubaa, A.; Mottola, L.; Zuniga, M.A.; Youssef, H.; Boano, C.A.; Alves, M. Radio link quality estimation in wireless sensor networks: A survey. *ACM Trans. Sens. Netw.* **2012**, *8*, 34:1–34:33.
2. Wang, X.; Bischoff, O.; Laur, R.; Paul, S. Localization in Wireless ad-hoc Sensor Networks using multilateration with RSSI for logistic applications. *Procedia Chem.* **2009**, *1*, 461–464.
3. Savvides, A.; Han, C.; Strivastava, M. Dynamic fine-grained localization in ad-hoc Networks of Sensors. In Proceedings of the International Conference on Mobile Computing and Networking, Rome, Italy, 16–21 July 2001; pp. 166–179.

4.   Benkic, K.; Malajner, M.; Planinsic, P.; Cucej, Z. Using RSSI value for distance estimation in wireless sensor networks based on ZigBee. In Proceedings of the 15th International Conference on Systems, Signals and Image Processing (IWSSIP'2008), Bratislava, Slovak Republic, 25–28 June 2008; pp. 303–306.

5.   Zhang, J.; Zhang, L. Research on distance measurement based on RSSI of ZigBee. In Proceedings of the International Colloquium on Computing, Communication, Control, and Management (CCCM'2009), Sanya, China, 8–9 August 2009; Volume 3, pp. 210–212.

6.   Liu, C.; Wu, K.; He, T. Sensor localization with ring overlapping based on comparison of received signal strength indicator. In Proceedings of the IEEE International Conference on Mobile Ad-hoc and Sensor Systems, Fort Lauderdale, FL, USA, 25–27 October 2004; pp. 516–518.

7.   He, T.; Huang, C.; Blum, B.M.; Stankovic, J.A.; Abdelzaher, T. Range-free localization schemes for large scale Sensor Networks. In Proceedings of the International Conference on Mobile Computing and Networking, San Diego, CA, USA, 14–19 September 2003, pp. 81–95.

8.   Bahl, P.; Padmanabhan, V.N. RADAR: An in-building RF-based user location and tracking system. In Proceedings of the Annual Joint Conference of the IEEE Computer and Communications Societies (INFOCOM'2000), Tel Aviv, Israel, 26–30 March 2000; pp. 775–784.

9.   Smailagic, A.; Kogan, D. Location sensing and privacy in a context-aware computing environment. *IEEE Wirel. Commun.* **2002**, *9*, 10–17.

10. Youssef, M.A.; Agrawala, A.; Shankar, A.U.; Noh, S.H. *A Probabilistic Clustering-Based Indoor Location Determination System*; Technical Reports; UMIACS: College Park, MD, USA, 2002.

11. Roos, T.; Myllymäki, P.; Tirri, H.; Misikangas, P.; Sievänen, J. A probabilistic approach to WLAN user location estimation. *Int. J. Wirel. Inf. Netw.* **2002**, *9*, 155–164.

12. Hassan-Ali, M.; Pahlavan, K. Site-specific wideband and narrowband modeling of indoor radio channel using ray-tracing. In Proceedings of the Personal, Indoor and Mobile Radio Communications (PIMRC'98), Boston, MA, USA, 8–11 September 1998; pp. 65–68.

13. Hassan-Ali, M.; Pahlavan, K. A new statistical model for site-specific indoor radio propagation prediction based on geometric optics and geometric probability. *IEEE Trans. Wirel. Commun.* **2002**, *1*, 112–124.

14. Awad, A.; Frunzke, T.; Dressler, F. Adaptive distance estimation and localization in WSN using RSSI measures. In Proceedings of the Conference Digital System Design Architectures, Methods and Tools (DSD'2007), Lubeck, Germany, 29–31 August 2007; pp. 471–478.

15. Vanheel, F.; Verhaevert, J.; Laermans, E.; Moerman, I.; Demeester, P. Automated linear regression tools improve rssi wsn localization in multipath indoor environment. *EURASIP J. Wirel. Commun. Netw.* **2011**, *2011*, 1–27.

16. Mihaylova, L.; Angelova, D.; Bull, D.R.; Canagarajah, N. Localization of Mobile Nodes in Wireless Networks with Correlated in Time Measurement Noise. *IEEE Trans. Mob. Comput.* **2011**, *10*, 44–53.

17. Patwari, N.; Hero, A.O.; Perkins, M.; Correal, N.S.; O'dea, R.J. Relative location estimation in wireless sensor networks. *IEEE Trans. Sig. Process.* **2003**, *51*, 2137–2148.

18. Itoh, K.-I.; Watanabe, S.; Shih, J.-S.; Sato, T. Performance of handoff algorithm based on distance and RSSI measurements. *IEEE Trans. Veh. Technol.* **2002**, *51*, 1460–1468.

19. Fischler, M.A.; Bolles, R.C. Random sample consensus: a paradigm for model fitting with applications to image analysis and automated cartography. *Commun. ACM* **1981**, *24*, 381–395.

20. De San Bernabé, A.; Martínez-de Dios, J.R.; Ollero, A. Efficient cluster-based tracking mechanisms for camera-based wireless sensor networks. *IEEE Trans. Mob. Comput.* **2015**, *14*, 1820–1832.

21. Sánchez-Matamoros, J.M.; Martínez-de Dios, J.R.; Ollero, A. Cooperative localization and tracking with a camera-based WSN. In Proceedings of the International Conference on Mechatronics (ICM2009), Malaga, Spain, 14–17 April 2009; pp. 1–6.

22. Jiménez-González, A.; Martínez-de Dios, J.R.; Ollero, A. An integrated testbed for heterogeneous mobile robots and other cooperating objects. In Proceedings of the IEEE/RSJ International Conference on Intelligent Robots and Systems, Taipei, Taiwan, 18–22 October 2010.

23. Jiménez-González, A.; Martínez-de Dios, J.R.; Ollero, A. An integrated testbed for cooperative perception with heterogeneous mobile and static sensors. *Sensors* **2011**, *11*, 11516–11543.

J. Sens. Actuator Netw. **2017**, *6*, 15

24. Rowe, A. CMUcam3 Datasheet. 2012. Available online: http://www.cmucam.org/ (accessed on 9 August 2017).
25. Fox, D.; Burgard, W.; Dellaert, F.; Thrun, S. Monte carlo localization: Efficient position estimation for mobile robots. In Proceedings of the AAAI/IAAI, Orlando, FL, USA, 18–22 July 1999; pp. 343–349.

Journal of
*Sensor and
Actuator Networks*

MDPI

Article

The Sensor Network Calculus as Key to the Design of Wireless Sensor Networks with Predictable Performance

Jens Schmitt [1,*], Steffen Bondorf [1] and Wint Yi Poe [2]

[1]   Distributed Computer Systems (DISCO) Lab, University of Kaiserslautern, 67663 Kaiserslautern, Germany;
     bondorf@cs.uni-kl.de
[2]   Huawei European Research Centre (ERC), 80992 Munich , Germany; wint.yi.poe@huawei.com
*   Correspondence: jschmitt@cs.uni-kl.de; Tel.: +49-631-205-3288

Received: 8 August 2017; Accepted: 5 September 2017; Published: 12 September 2017

Abstract: In this article, we survey the sensor network calculus (SensorNC), a framework continuously developed since 2005 to support the predictable design, control and management of large-scale wireless sensor networks with timing constraints. It is rooted in the deterministic network calculus, which it instantiates for WSNs, as well as it generalizes it in some crucial aspects, as for instance in-network processing. Besides presenting these core concepts of the SensorNC, we also discuss the advanced concept of self-modeling of WSNs and efficient tool support for the SensorNC. Furthermore, several applications of the SensorNC methodology, like sink and node placement, as well as TDMA design, are displayed.

Keywords: network calculus; wireless sensor networks; performance analysis

1. Introduction

Many applications of wireless sensor networks (WSN) require timely actuation. For example, industrial process automation typically consists of a multitude of sensors and actuators that are required to interact in a very clearly-defined manner with respect to their timing. Traditionally, in this environment, real-time conditions have to be met. While using wireless communications and the increased complexity of modern factories makes hard real-time a more elusive goal, there is still a need for a predictable timing behavior of the system in order to avoid catastrophic behavior. Furthermore, WSNs have been proposed in the context of emergency response systems. Again, it is obvious that predictable timing behavior is key to the acceptance of such systems. More generally, many envisioned applications of cyber-physical systems (CPS) typically can be viewed as closed-loop systems [1], in terms of control theory, such that the sensing often becomes time-critical in order to ensure the stability of the system. As in many of the visions of CPS, using WSNs is an integral part for the sensing requirements, it can hardly be overemphasized that predictable timing is a necessity for WSNs in that context, as well. Consequently, a mathematical methodology to dimension, operate and control the timing behavior in WSNs is of outmost importance.

There is a clear trend for WSNs to become of ever larger scale, and some examples can be found in the CitySee project [2], the GreenOrbs project [3] and more generally in the emergence of the Internet of Things (IoT), which partially can be seen as a world-wide sensor network (with many sensors using wireless communication). Hence, a second key criterion for WSNs is the scalability of its basic functions and, in particular, of the mathematical framework to analyze its timing behavior. To that end, there is a clear need for a fast mathematical methodology to predict the timing behavior in WSNs.

In 2005, the sensor network calculus (SensorNC) was proposed as such a mathematical methodology [4], and in more than a decade, it was developed to meet the special requirements of time-sensitive, large-scale WSNs. Mathematically, it can be viewed as a special instance of the general network calculus, which itself is based on a min-plus algebraic formulation for the performance analysis of queueing networks as typical in packet-switched networks [5,6].

The goal is to have a mathematical framework that allows one to compute useful performance characteristics of WSNs such as message transfer delays, required buffer sizes and link capacities, but also duty cycle durations, to name some of the most prominent ones. Sometimes, we may require absolute values for, e.g., delays, if a certain application has hard real-time requirements, but sometimes, it may also suffice to have a representative relative metric in order to compare different design alternatives of a WSN before its deployment, for instance.

To that end, the SensorNC was customized in several dimensions and also extended over the general network calculus to capture the special requirements of WSNs, e.g., in-network processing. The goal of this article is to provide an overview of the efforts in over a decade of the development of the SensorNC framework, emphasizing the most important milestones in this.

In the following section, some background on the general network calculus framework is provided, in order to alleviate the introduction of the basic SensorNC methodology in Section 3. Advanced concepts of the SensorNC, in particular, in-network processing, are presented and illustrated in Section 4. Section 5 provides an overview of the self-modeling capabilities of WSNs that employ SensorNC. In Section 6, the important aspect of tool support for the SensorNC is discussed. Clearly, a mathematical framework such as the sensor is only as useful as the problems it can solve, and thus, we present several examples of SensorNC applications in Section 7. Section 8 provides a discussion and some concluding remarks.

2. Background on Network Calculus

We start with the necessary background on network calculus, before we introduce its customization in the WSN context in the following section.

2.1. Modeling of Flows and Performance Characteristics

Network calculus models the sequence of packets that define a flow's data arrivals as non-negative, wide-sense increasing functions. These cumulatively count arriving data over time:

$$\mathcal{F}_0 = \left\{ f : \mathbb{R}^+ \to \mathbb{R}^+ \mid f(0) = 0,\ \forall s \leq t \ :\ f(t) \geq f(s) \right\}.$$

A flow's input, as well as its output from a system $\mathcal{S}$ are of interest to performance modeling. We denote the input up to time t as $R(t)$ and the output as $R^*(t)$. If both count the data of the same flow, we demand $\forall t \in \mathbb{R}^+ \ :\ R(t) \geq R^*(t)$, i.e., the flow's output from system $\mathcal{S}$ is caused by the input to $\mathcal{S}$. This causality preservation is known as the flow constraint of network calculus.

The flow definition and its causal transformation from system input to output enables us to define first the performance characteristics.

Definition 1. *(Backlog and delay) Assume a flow with input function R traverses a system $\mathcal{S}$ and results in the output function R*. The backlog of the flow at time t is defined as:*

$$B(t) = R(t) - R^*(t).$$

The (virtual) delay for a data unit arriving at $\mathcal{S}$ at time t is defined as:

$$D(t) = \inf \left\{ \tau \geq 0 \mid R(t) \leq R^*(t + \tau) \right\}.$$

2.2. Network Calculus Performance Analysis

Network calculus operates on bounding functions for flow arrivals. These are derived from the above input functions; yet, they are not defined over the time t that passed until the observation of the flow. Instead, they are defined over the duration of an observation d. These bounding functions are called arrival curves.

Definition 2. *(Arrival curve) Given a flow with input function R, a function $\alpha \in \mathcal{F}_0$ is an arrival curve for R iff:*

$$\forall 0 \leq d \leq t : R(t) - R(t-d) \leq \alpha(d)$$

SensorNC often restricts the set of arrival curves to token-bucket-shaped traffic:

$$\mathcal{F}_{TB} = \left\{ \gamma_{r,b} : \mathbb{R}^+ \to \mathbb{R}^+ \mid \gamma_{r,b}(0) = 0, \underset{d>0}{\forall} \gamma_{r,b}(d) = b + r \cdot d \right\} \subseteq \mathcal{F}_0, \, r, b \geq 0.$$

Curves of $\mathcal{F}_{TB}$ can easily be applied to bound the typical behavior of wireless sensor nodes. Assume a node periodically measuring its environment to report data. The maximum size of a measurement translates to parameter b that bounds the flow's burstiness. The bound on the subsequent data reporting rate is $r = \frac{b}{p}$ where p denotes the sensing period.

Analyzing a network requires transformations of the bounding functions of (sensor) network calculus. Operations in $(\wedge, +)$-algebra have been established; see [5–7], respectively. The most important operations of this $(\wedge, +)$-algebraic framework over $\mathcal{F}_0$ are presented in the following.

Definition 3. $(\wedge, +)$-*operations) The $(\wedge, +)$-algebraic aggregation, convolution and deconvolution of two functions $f, g \in \mathcal{F}_0$ are defined as:*

$$\text{aggregation: } (f + g)(d) = f(d) + g(d),$$
$$\text{convolution: } (f \otimes g)(d) = \underset{0 \leq s \leq d}{\inf} \{f(d-s) + g(s)\},$$
$$\text{deconvolution: } (f \oslash g)(d) = \underset{u \geq 0}{\sup} \{f(d+u) - g(u)\}.$$

Note that deconvolution is not exactly dual to convolution [6].

Aggregation of flows that cross a common system requires aggregating their arrival curves. In SensorNC, the flow aggregate's arrival curve can be computed easily if each individual flow is constrained by a token-bucket arrival curve.

Corollary 1. *(Arrival curve aggregation in SensorNC) For the aggregation of n arrival curves of $\mathcal{F}_{TB}$, it holds that:*

$$\sum_{i=1}^{n} \gamma_{r_i, b_i} = \gamma_{\sum_{i=1}^{n} r_i, \sum_{i=1}^{n} b_i}.$$

In addition to flows, network calculus also operates on functions bounding the service capabilities of a system S. These so-called service curves are defined over the duration of observation, as well. They bound the worst-case transformation of an input function to an output function.

Definition 4. *(Service curve) If the service provided by a system S for a given input function R results in an output function R*, we say that S offers a service curve β iff:*

$$R^* \geq R \otimes \beta.$$

In SensorNC, service is often characterized by rate-latency functions from:

$$\mathcal{F}_{RL} = \{\beta_{R,T} : \mathbb{R}^+ \to \mathbb{R}^+ \mid \beta_{R,T}(d) = \max\{0, R \cdot (d-T)\}\} \subseteq \mathcal{F}_0, \, T \geq 0, \, R > 0.$$

Rate-latency functions $\beta_{R,T} \in \mathcal{F}_{RL}$ can model TDMA channel access [8] and duty cycling [9] in wireless sensor networks.

When data are queued to be forwarded and service capacity is available, many systems guarantee a greater amount of service than the one of Definition 4. Thus, they fulfil a stricter definition of service curves. These service curves allow for some computations that those from the general service curve model of Definition 4 do not.

Definition 5. *(Strict service curve) Let $\beta \in \mathcal{F}_0$. System $\mathcal{S}$ offers a strict service curve β to a flow if, during any backlogged period of duration d, the output of the flow is at least equal to $\beta(d)$.*

Network calculus often computes results for one specific flow. In cases where this flow competes for a system's resources with other flows, a bound on its resource share needs to be derived. This left-over service curve can only be computed from a strict service curve.

Theorem 1. *(Left-over service curve) Consider a system $\mathcal{S}$ that offers a strict service curve β. Let $\mathcal{S}$ be crossed by two flows R_1, R_2 with arrival curves α_1 and α_2, respectively. Then, R_1's worst-case residual service share under arbitrary multiplexing at system $\mathcal{S}$, i.e., the left-over service curve for the flow of interest, is:*

$$\beta^{l.o.} = \beta \ominus \alpha_1 := \sup_{0 \le u \le d} \{(\beta - \alpha)(u)\}.$$

As stated above, service curves bound the transformation of an input function $R(t)$ to an output function $R^*(t)$. Given an arrival curve for $R(t)$, they can also be applied to compute an arrival curve for $R^*(t)$, a so-called output arrival curve:

Theorem 2. *(Output arrival curve) Assume a flow f has an arrival curve α, and consider f traversing the system $\mathcal{S}$ offering a service curve β. After being transformed by $\mathcal{S}$, i.e., at the system's output, f is bounded by the arrival curve:*

$$
\begin{aligned}
\alpha^*(d) \;&=\; (\alpha \oslash \beta)(d). \\
&=\; \begin{cases} 0 & \text{if } d = 0 \\ (\alpha \oslash \beta)(d) & \text{otherwise} \end{cases}.
\end{aligned}
$$

Note that deconvolution is not closed in $\mathcal{F}_0$ as $(\alpha \oslash \beta)(0) \ge 0$. Therefore, we slightly augmented the operation such that its result is guaranteed to pass through the origin. Other performance bound computations are not affected by this adaptation.

Theorem 3. *(Performance bounds) Consider a system $\mathcal{S}$ that offers a service curve β. Assume a flow f with arrival curve α traverses the system. Then, we obtain the following performance bounds for f:*

$$\text{backlog: } \forall t \in \mathbb{R}^+ : \; B(t) \le (\alpha \oslash \beta)(0),$$
$$\text{delay: } \forall t \in \mathbb{R}^+ : \; D(t) \le \inf \{d \ge 0 \mid (\alpha \oslash \beta)(-d) \le 0\}.$$

Last, we present a central result of network calculus: the concatenation theorem. Concatenation allows one to logically transform a tandem of systems into a single system whose capabilities are bounded by a single service curve.

Theorem 4. *(Concatenation theorem for tandem systems) Consider a flow that traverses a tandem of systems $\mathcal{S}_i, i = 1, \ldots, n$. Each $\mathcal{S}_i$ offers a service curve $\beta_{\mathcal{S}_i}$ to the flow. Then, the concatenation of the n systems offers a service curve $\bigotimes_{i=1}^{n} \beta_{\mathcal{S}_i}$ to the flow.*

A system-by-system analysis of delay and backlog bounds that applies Theorems 2 and 3 does not guarantee tight results. In contrast, the concatenation theorem facilitates tightness of bounds by deriving the tandem's end-to-end service curve for the flow crossing it. In particular, bounds scale linearly in the number of crossed systems. This phenomenon is known as pay bursts only once [6].

An overview of the network calculus notation is provided in Table 1.

Table 1. Network calculus notation for functions of arrivals and service, as well as their min-plus algebraic manipulation and performance bounds.

Quantifier	Definition
$\mathcal{F}_0$	Non-negative, wide-sense increasing functions passing through the origin
$\gamma_{r,b} \in \mathcal{F}_{TB}$	Token-bucket functions with bucket size b and rate r
$\beta_{R,T} \in \mathcal{F}_{RL}$	Rate-latency functions with rate R and latency T
R_i	Input function of the flow originating at node i
$\bar{R}_i$	Aggregate input function for all flows at node i
R_i^*	Aggregate output function for all flows at at node i
α, α^f	Arrival curve, arrival curve of flow f
α_i	Arrival curve of the flow originating at node i
$\bar{\alpha}_i$	Aggregate arrival curve for all flows at node i
α_i^*	Aggregate output arrival curve for all flows at node i
β, β_i	Service curve, service curve of node i
$\beta^{l.o.}, \beta_i^{l.o.}$	Left-over service curve, left-over service curve of node i
$\beta_i \otimes \beta_j$	Service curve concatenation with min-plus convolution $\otimes$
$\alpha \oslash \beta$	Output bound computation with (adapted) min-plus deconvolution $\oslash$
$\beta \ominus \alpha$	Left-over service curve computation with non-decreasing subtraction
$D(t), D_i$	Delay at time t and the time-invariant delay bound at node i
$B(t), B_i$	Backlog at time t and the time-invariant backlog bound at node i

3. The Sensor Network Calculus

Provision of worst-case bounds on performance measures, such as the maximum delay that any flow in a WSN experiences, can be achieved by the sensor network calculus (SensorNC) framework. It was first presented in [4]. This work constitutes the first step towards concise worst-case analysis of WSNs. The initial SensorNC was based on the analysis of single servers in isolation; end-to-end performance bounds were derived by the addition of the crossed servers' bounds. The concatenation theorem from conventional network calculus and thus a holistic view on the WSN was not applied (yet). This is due to the fact that the typical sink tree structure of the network does establishes an interference pattern of flows that does not allow for a direct application of the concatenation theorem. Additively-derived results are known to be more pessimistic and so are the performance bounds. Later work on SensorNC was then targeted to taking advantage of the concatenation result.

3.1. Sensor Network System Model

WSNs are often oriented towards a single base station such that the employed routing protocol can form a sink tree. This commonly-found class of operations can be abstractly modeled as shown in Figure 1. Only traffic from sensors to the base station is explicitly taken into account in this model. The majority of traffic flows in this direction. Traffic in the opposite direction, e.g., topology control messages or node configurations, can be considered in the service curves for sensor-to-sink communication; for instance, by assuming strict priority. Overall, the network calculus model arranges the sensor nodes as servers in a directed acyclic graph as depicted in Figure 1.

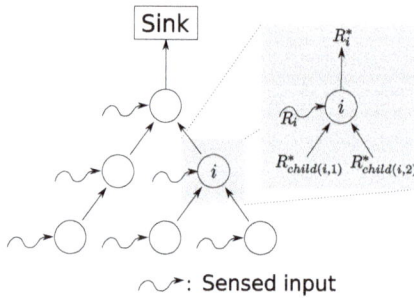

: Sensed input

Figure 1. Sensor network model.

The system model takes the following aspects into account: each sensor senses its environment, reports data to the sink and forward measurements generated by other sensors in the network, i.e., the traffic seen by any sensor i consists of an input function R_i derived from its own sensing and, in case it is not a leaf node in the sink tree, the forwarded data from its child nodes $child\,(i,1), ..., child\,(i,n_i)$, where n_i denotes the number of these nodes connected to sensor node i. As all of this input data are forwarded by sensor i, an output function R_i^* results. This output function is processed by sensor i's parent node.

3.2. Incorporation of Network Calculus Components

In this section, we incorporate the basic network calculus components, i.e., arrival curves and service curves, into the network system model. We make use of the simplicity of the SensorNC sink tree where each sensor has exactly one parent node. For a detailed treatment of modeling generic feed-forward networks, see [10].

Each sensor in the network is crossed by at least one flow. Their aggregate defines the total input to sensor i, $\bar{R}_i$. The total input function is composed by the sensor's sensed input and the output of its child nodes:

$$\bar{R}_i = R_i + \sum_{j=1}^{n_i} R_{child(i,j)}^*. \tag{1}$$

We can derive an analogous description with network calculus arrival curves and operations applied to them:

$$\bar{\alpha}_i = \alpha_i + \sum_{j=1}^{n_i} \alpha_{child(i,j)}^*. \tag{2}$$

where $\bar{\alpha}_i$ denotes the arrival curve for $\bar{R}_i$. It is composed of the arrival curve for the input sensed by sensor i, α_i, and the aggregation of its child nodes' output arrival curves. As an example, we could use simple token-bucket functions to model these inputs.

Next, service curves need to be incorporated. Specifying them is subject to the packet scheduling applied by the respective server. In a WSN, an additional impact factor is the link layer characteristics. For instance, duty cycling may be applied to achieve predefined energy-efficiency targets and thus impacts the service curve by periodical unavailability of the medium access. We model such a periodic service curve in the following. Assume the full medium capacity C is periodically available after an initial delay T, i.e., a TDMA medium access pattern results after this delay (see Figure 2). Let the TDMA frame duration be f and assume a sensor node receives s time units of service within a frame. From this information, we can compute T, the maximum initial medium access latency that can be experienced, as $T = f - s$. To establish independence from the start of observation, the sensor node cannot be assumed to receive service during an initial duration of observation of T, and the service curve is zero. Then, the node receives service for the duration of its share of the TDMA frame. That is, from T to $T + s$, the service curve will have a linear segment of slope C until it reaches sC. For the remainder of the TDMA frame, i.e., from $T + s$ to $T + f$, service is unavailable again, and the service curve remains at sC. With the start of the next TDMA frame, the shape of the service curve from T to $T + f$ is repeated; it actually repeats indefinitely with a period of f. A similar service curve structure was proposed for 802.15.4 networks [11]. This curve can be approximated with a rate-latency service curve consisting of only two linear segments [12,13]: $\beta_{R,T}(t) = \max\left(R(t - T), 0\right)$ with $R = \frac{s}{f}C$ and $T = f - s$. In Figure 2, this fluid version of the TDMA service curve is labeled β.

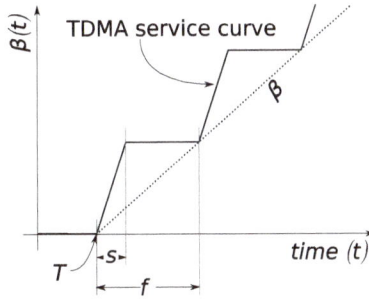

Figure 2. TDMA service curve.

3.3. Calculation of Network-Internal Traffic Flow

When incorporating the required network calculus components into the system model, we added output arrival curves of servers. Next, we present their computation in a sink tree network. The traffic a sensor node i forwards to its parent sensor node is constrained by arrival curve:

$$\alpha_i^* = \tilde{\alpha}_i \oslash \beta_i. \tag{3}$$

Equation (3) can be expanded to a recursive computation by considering the output of child nodes as given in Equation (2). However, making use of the sink tree structure of the network allows for the iterative computation of all α_i^* given by Algorithm 1. Note that this procedure requires perfect knowledge about the sink tree structure to correctly handle aggregation of flows in Step 3.

Algorithm 1: Computation of network-internal traffic.

---

1. Assume a sink tree network consisting of n sensor nodes; all arrival curves α_i are given at the locations sensed input enters the network and service curves β_i are known for all sensor nodes, $i = 1, \ldots, n$.
2. The output arrival curve computation for leaf nodes strictly follows Theorem 2:
   (a) $\alpha_i^* = \alpha_i \oslash \beta_i$ where sensor i is a leaf node.
   (b) Mark every sensor i as "calculated".
3. Continue with a non-leaf node i whose child nodes are all marked "calculated", but is not marked itself. We apply Equation (3):
   (a) $\alpha_i^* = \left(\alpha_i + \sum_{j=1}^{n_i} \alpha_{child(i,j)}^*\right) \oslash \beta_i$ where $child(i, j)$ are the n_i child nodes of non-leaf sensor i that were marked "calculated".
   (b) Mark non-leaf sensor i as "calculated".
4. Check if all sensor nodes just below the sink are marked "calculated".
   (a) Yes: the algorithm terminates.
   (b) No: repeat Step 3.

3.4. Calculation of Performance Bounds

After the arrival curves for network-internal traffic flows are computed, we can also compute performance bounds according to Theorem 3. For instance, the buffer requirements of sensor nodes B_i, a performance metric of interest for the bottleneck nodes directly below the sink:

$$B_i = v(\tilde{\alpha}_i, \beta_i) = \sup_{s \geq 0}\{\tilde{\alpha}_i(s) - \beta_i(s)\}, \tag{4}$$

Similarly, the per-node delay bounds D_i are computed with the given arrival and service curves:

$$D_i = h(\tilde{\alpha}_i, \beta_i) = \sup_{s \geq 0}\{\inf\{\tau \geq 0 : \tilde{\alpha}_i(s) \leq \beta_i(s + \tau)\}\}. \tag{5}$$

These derivations constitute SensorNC's original node-by-node computation of performance bounds [4]. In order to compute a flow's end-to-end delay bound, the per-node delay bounds along its path are added up. This approach is known as the total flow analysis (TFA) of network calculus. It does not benefit from the pay bursts only once (PBOO) phenomenon.

A more sophisticated approach is required to benefit from more advanced network calculus phenomena and thus derive more accurate end-to-end delay bounds. PBOO can be achieved by first computing the left-over service curve for every server on the flow of interest's path (application of Theorem 1). Then, these left-over service curves are concatenated with Theorem 4 before the delay bound is computed. This approach is known as the separated flow analysis (SFA) of network calculus and its application in SensorNC was proposed in [11] where the improvement over TFA is also given.

The third phenomenon to tighten end-to-end delay bounds in network calculus is called pay multiplexing only once (PMOO). In case cross-traffic flows are multiplexed with the flow of interest on multiple consecutive nodes (as is common in sink trees), it prevents their burstiness from impacting the derivation multiple times. This is achieved by applying the concatenation theorem as early as possible, i.e., before the application of Theorem 1.

We illustrate the impact of advances in the SensorNC analysis by an example. Consider the simple sink tree network depicted in Figure 3. Computing delay bounds proceeds as discussed above: TFA computes an additive end-to-end delay bound; SFA computes per-node left-over service curves; and PMOO concatenates before the left-over computation:

$$
\begin{aligned}
D^{\text{TFA}} &= h\left(\alpha_1 + \alpha_2, \beta_1\right) + h\left((\alpha_1 + \alpha_2) \oslash \beta_1, \beta_2\right) \\
D^{\text{SFA}} &= h\left(\alpha_1, [\beta_1 \ominus \alpha_2] \otimes [\beta_2 \ominus (\alpha_2 \oslash \beta_1)]\right) \\
D^{\text{PMOO}} &= h\left(\alpha_1, [(\beta_1 \otimes \beta_2) \ominus \alpha_2]\right)
\end{aligned}
$$

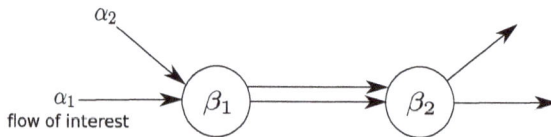

Figure 3. Simple example to illustrate bounding methods.

Assume rate-latency service curves $\beta_1 = \beta_2 = \beta_{3,0}$ and token-bucket arrival curves $\alpha_1 = \alpha_2 = \gamma_{1,1}$. Then, we compute the following delay bounds: $D^{\text{TFA}} = \frac{4}{3}$, $D^{\text{SFA}} = \frac{3}{2}$, $D^{\text{PMOO}} = 1$. As expected, PMOO results in the most accurate bound. In our example, TFA outperforms SFA, yet only because it implicitly assumes FIFO multiplexing (when computing the horizontal deviation for the total flow aggregate at each node) instead of the more pessimistic arbitrary multiplexing assumption of SFA and PMOO. Nevertheless, in larger sink tree networks, application of the SensorNC SFA usually outperforms the TFA [11].

More details about the PMOO analysis for general feed-forward networks, including its relation to other network calculus analyses, as well as an in-depth discussion, can be found in [14,15]. In sink tree networks, a specialized version can make use of the network structure. Similar to Algorithm 1, perfect knowledge about the locations of aggregation of flows and the lack of demultiplexing due to their common sink are exploited in Algorithm 2.

Algorithm 2: Sink tree PMOO analysis.

---

1.   Let $M = \{1, \ldots, k\}$ be the set of nodes, each providing a service curve β_i; a flow of interest is traversing on the way from its source to the sink. The interfering traffic, from the perspective of the flow of interest, at node i is denoted as α_i (see Section 3.3).
2.   Let $\beta_{k+1}^{\text{l.o.}} = \delta_0$ with:

$$\delta_d(t) = \begin{cases} 0 & \text{if } t \leq d \\ \infty & \text{otherwise} \end{cases}$$

   where δ_0 is the neutral element of the min-plus convolution.
3.   Starting from the sink node k, going to Node 1, the source node of the flow of interest calculate left-over service curves for node i as:

$$\beta_i^{\text{l.o.}} = \left(\beta_{i+1}^{\text{l.o.}} \otimes \beta_i \right) \ominus \alpha_i.$$

4. $\beta_1^{\text{l.o.}}$ is the left-over service curve for the flow of interest.

---

4. Advanced SensorNC: In-Network Processing

In the previous section, sensor nodes were abstracted as simple communication resources. Clearly, this can only be a very coarse-grained abstraction. In many wireless sensor networks, some form of in-network processing is applied to the data before it is delivered towards the sink. This is, e.g., often done in order to save energy by data aggregation. Therefore, computational resources, i.e., the usage of the processing unit, should be factored into the SensorNC model of nodes. As the workload units differ between communication and computation, we need new modeling elements that translate between the usage of communication and computational resources. In [16,17], such elements have been introduced as workload transformations or scaling elements, respectively. Though being conceptually very simple components that translate, for instance, a number of bytes received from other sensors into a worst-case sequence of processing steps, they complicate the end-to-end analysis since they build up "walls" between the different resources, effectively inhibiting the direct usage of the concatenation result. Nevertheless, we show that an end-to-end-analysis can still be performed by moving the scaling elements such that simultaneously, the analysis becomes feasible and no compromise of the worst-case occurs.

4.1. Background on Data Scaling in Network Calculus

Next, we introduce the necessary background on scaling elements as presented in [17].

Definition 6. *(Scaling function) A scaling function $S \in \mathcal{F}$ assigns an amount of scaled data $S(a)$ to an amount of data a.*

Scaling functions are a very general concept and serve as a model for any kind of data transformation in a network calculus model. Note that they do capture any queueing-related effects;

consequently, scaling is assumed to be done infinitely quickly. Queueing effects are captured by the service curve element.

Definition 7. *(Scaling curves) Given a scaling function S, two functions $\underline{S}, \overline{S} \in \mathcal{F}$ are minimum and maximum scaling curves of S iff $\forall b \geq 0$; it applies that:*

$$\underline{S}(b) \leq \inf_{a \geq 0} \{S(b+a) - S(a)\}$$

$$\overline{S}(b) \geq \sup_{a \geq 0} \{S(b+a) - S(a)\}$$

In [17], it is shown that maximum scaling curves should be sub-additive and minimum scaling curves super-additive. Otherwise, they can be enhanced by computing their super-additive resp. sub-additive closure.

Theorem 5. *(Alternative systems) Consider the two systems in Figure 4, and let $R(t)$ be the input function. System (a) consists of a server with minimum service curve β whose output is scaled with scaling function S, and System (b) consists of a scaling function whose output is input to a server with minimum service curve β_S. Given System (a), the lower bound of the output function of System (b), R_S^*, that is $S(R) \otimes \beta_S$, is also a valid lower bound for the output function of System (a) if:*

$$\beta_S = \underline{S}(\beta).$$

The implication of this theorem is that performance bounds for System (b) are also valid bounds for System (a), that is a scaling element can be moved in front of a service curve element if we apply the minimum scaling curve to the service curve of the respective component.

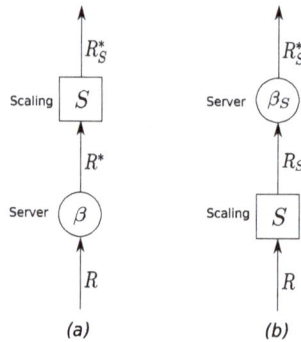

Figure 4. Alternative systems.

The effect of scaling on the arrival constraints of a flow is stated in the following corollary.

Corollary 2. *(Arrival constraints under scaling) Let R be an input function with arrival curve α that is fed into a scaling function with maximum scaling curve $\overline{S}$. An arrival curve for the scaled output from the scaling element is given by:*

$$\alpha_S = \overline{S}(\alpha).$$

Not that, if the arrival curve α and maximum scaling curve $\overline{S}$ are tight, then the scaled arrival curve $\overline{S}(\alpha)$ also is.

4.2. Data Scaling in Sink Trees

With the aid of the scaling element, a much more comprehensive model of a WSN that integrates the processing besides the communication resources can be facilitated. Figure 5 illustrates the advanced SensorNC model.

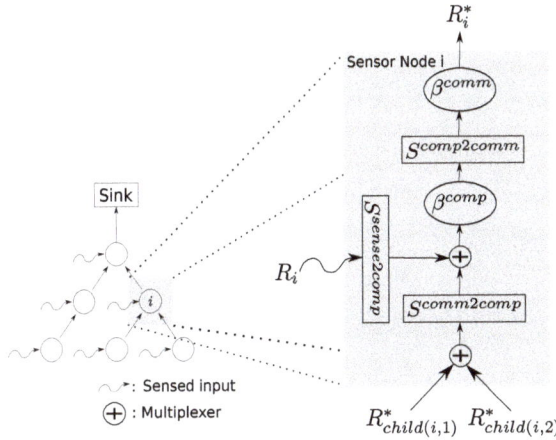

Figure 5. Advanced sensor network model.

Besides the scaling elements, a multiplexing element is also introduced. The multiplexing element enables an explicit modeling of the aggregation of a set of flows, for instance arbitrary, FIFO or using strict priorities between flows. Thus, data flows arriving from predecessors in the sink tree are multiplexed and then scaled in order to transform them into local processing demand. The local sensing data are taken into account by a scaling element just before they are multiplexed with the other data flows. The overall aggregate is then processed by the sensor node modeled by a service curve β^{comp}. As the amount of data that are forwarded downstream depends on how the processing is, for example, able to compress the data flow by, e.g., data aggregation, another scaling element translates the flow back to the required communication resources before it is served by the communication subsystem of the sensor node represented by another service curve element β^{comm}.

The challenge of this new advanced SensorNC model lies in the scaling elements that prevent an immediate end-to-end analysis following the PMOO principle. Yet, building on the theory from the previous subsection, we can shift all of the scaling elements upstream across service curve elements in order enable a true (PMOO) analysis again. The only remaining challenge is the shifting of scaling elements across multiplexers. In the following theorem, we prove a sufficient condition that allows this shifting without compromising the worst-case semantics.

Theorem 6. *(Shift of the scaling element across the multiplexer) Assume a situation as depicted in Figure 6, i.e., two flows are multiplexed and then fed into a scaling function S with maximum scaling curve $\overline{S}$ and minimum scaling curve $\underline{S}$. Provided that the minimum scaling curve is super-additive and the maximum scaling curve is sub-additive, we can transform System (a) into System (b) without improving the worst-case scenario in System (b) over the one in System (a).*

Proof. This can be found in [18]. □

Consequently, all of the scaling elements can be moved to the sources of traffic. In principle, all of the input to the system is translated to the same resource units before entering a system of "homogenized" servers, such that we can apply the end-to-end concatenation-based techniques from Section 3.

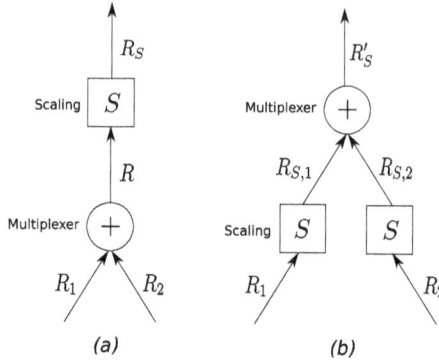

Figure 6. Scaling element with multiplexer.

4.3. Effect of Scaling Element Movement on Model Components

To make the method of scaling element movements more tangible, let us briefly illustrate its effect on the other modeling components. We base this discussion on simple, yet very commonly-used modeling component instances. In particular, we assume token-bucket arrival curves $\gamma_{r,b} \in \mathcal{F}_{TB}$ as arrival curves and rate-latency service curves $\beta_{R,T} \in \mathcal{F}_{RL}$. As the maximum scaling curve, we also assume a token-bucket function γ_{r_S,b_S}, and as minimum scaling curve, we take a rate-latency curve β_{R_S,T_S}. This form of maximum scaling curve is sensible as it can capture the fact that a small number of bytes received might result in a high demand on the computational resources, whereas for larger amounts of received data, this effect degrades. For the minimum scaling curve, the reasoning is just the other way around. Now, when we shift the scaling element across a service curve element, we need to compute the new service curve as:

$$\underline{S}(\beta) = \beta_{R_S,T_S} \circ \beta_{R,T} = \beta_{R_S R,T + \frac{T_S}{R}},$$

where $\circ$ denotes the concatenation of functions. Hence, the resulting service curve is again of the rate-latency type with a suitably scaled rate and an increased latency.

If the scaling elements have been shifted to the sources of traffic, i.e., the sensing inputs, their effect on the arrival constraints can be calculated as:

$$\overline{S}(\alpha) = \gamma_{r_S,b_S} \circ \gamma_{r,b} = \gamma_{r_S r, b_S + r_S b}$$

Thus, the arrival curves of the scaled traffic flows remain token buckets with again suitably-scaled rate and increased bucket depths.

Moving scaling elements across multiplexers has been discussed in the previous section and does not alter the multiplexers. That is true for any multiplexing, be it arbitrary, FIFO, priority-based or any other.

4.4. Numerical Experiment

In this section, we perform a set of numerical experiments to investigate the performance benefits of a holistic SensorNC analysis using the PMOO result and the shifting of scaling elements from Section 4.2 compared to a node-by-node approach based on the total flow analysis (TFA) from Section 3.4.

4.4.1. Experimental Design

At first, we create a set of WSN topologies by randomly placing n sensor nodes on a square field using a uniform random distribution. All nodes have a transmission range of 20 m such that connectivity is likely achieved (if not, the corresponding topologies are discarded). The sink node is in the middle of the sensor field, and the sink tree is built from a shortest path algorithm. For statistical significance, we evaluate 10 topologies for each experiment scenario. In all experiments, we selected $n = 100$ nodes and a sensor field of 100 m^2.

The parameters relating to the sensor nodes resemble MICAz motes from Crossbow Technology Inc. (Milpitas, CA, USA) [19]. A transmission rate of 250 kbps is used; as MAC, we assumed a TDMA-based scheme that uses a duty cycle of 1% and a TDMA frame length of 100 ms. Thus, the communication service curve is given as $\beta^{comm} = \beta_{2.5[kbps],0.099[s]}$. As the packet length, 36 B TinyOS packets are assumed. With respect to computational resources, the MICAz has an Atmel ATmega 128 L microcontroller operating at 8 MHz. Assuming an average of four cycles per instruction, its raw capacity amounts to 2 MIPS. Under the realistic assumption that the processor also performs other tasks, for instance network control operations, and that a power management scheme with certain sleep periods in low power modes is applied, we select its capacity for data processing to be 10% of its full capacity with a worst-case latency of 1 ms. Hence, the service curve for computation is given by $\beta^{comp} = \beta_{0.2[MIPS],0.001[s]}$.

From the scaling elements, we employ token-bucket functions for the maximum scaling curves and rate-latency functions for the minimum scaling curves. The maximum scaling curve that translates between communication and computation resources is set as $\overline{S}^{comm2comp} = \gamma_{5000[Instr./p],b[Instr.]}$, that is a received packet results in 5000 instructions, and deviations from this are captured by the bucket depth b, which varies in the experiments. The minimum scaling curve is set to $\underline{S}^{comm2comp} = \beta_{5000[Instr./p],T[p]}$, with the latency parameter T also being varied in the experiments. Along the same lines, maximum and minimum scaling curves for the translation between sensing and computation resources are set as $\overline{S}^{sense2comp} = \gamma_{6000[Instr./p],b[Instr.]}$ and $\underline{S}^{sense2comp} = \beta_{6000[Instr./p],T[p]}$; here a higher computational demand for sensing is assumed because raw sensing values are given priority. Rescaling computation into communication resource demand is selected to be roughly inverse to the other scaling elements while also capturing some compression because of, e.g., data aggregation.

The data arrival from local sensor are modeled by a token bucket $\gamma_{0.1[p/s],1[p]}$, that is we assume a packet to be created every 10 s with a burst due to an instantaneous packet arrival.

Clearly, many of these settings seem somewhat arbitrary and differ from scenario to scenario, yet they are from realistic settings and should thus be fairly representative.

With the aid of the DISCO Deterministic Network Calculator [20,21], we next present the results from the comparison between holistic and component-wise analysis (the DISCO Deterministic Network Calculator is publicly available under https://disco.cs.uni-kl.de/index.php/projects/disco-dnc.)

4.4.2. Benefit of End-To-End Analysis

At first, in a baseline comparison, minimum and maximum scaling curves are assumed to be identical and pass through the origin. In Figure 7, we can observe the SensorNC outcomes in terms of delay bounds for the holistic and component-wise methods for 10 different random topologies with 100 nodes each.

The holistic PMOO is superior to the TFA, which results in delay bounds up to 1.9-times higher than for PMOO.

Sometimes, we cannot assume identical minimum and maximum scaling curves as in the previous experiment because of non-determinisms in the actual processing. However, often, we can use upper and lower bounds on the amount of processing that is required, i.e., we can model this by differing minimum and maximum scaling curves. However, this bounding may adversely effect the performance of the PMOO analysis method. In order to provide more insights into this, we next change the minimum and maximum scaling curves to be no longer passing through the origin, but move the maximum scaling curve a bit upwards and shift the minimum scaling curves to the right. Therefore, they are drifting apart from each other, i.e., we set the bucket depth parameters of the maximum scaling curves, as well as the latency parameters of the minimum scaling curves incrementally higher until certain limits are reached. These limits are: for $\overline{S}^{comm2comp}$, the maximum bucket depth is 240; for $\underline{S}^{comm2comp}$, the maximum latency is 0.006; for $\overline{S}^{sense2comp}$, the maximum bucket depth is 300; for $\underline{S}^{sense2comp}$, the maximum latency is 0.006; for $\overline{S}$, the maximum bucket depth is 0.006; for $\underline{S}^{comp2comm}$, the maximum latency is 240.

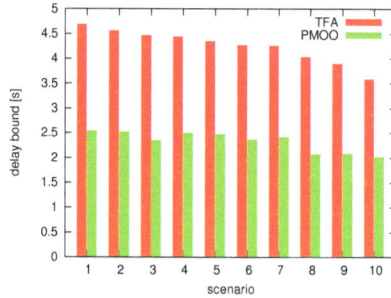

Figure 7. Baseline comparison between pay multiplexing only once (PMOO) and total flow analysis (TFA) for different scenarios.

In Figure 8, for both methods, we can observe the average delay bounds for 10 random topologies under different expansion factors, with the expansion factor as the percentage of how much the drifting limits have used.

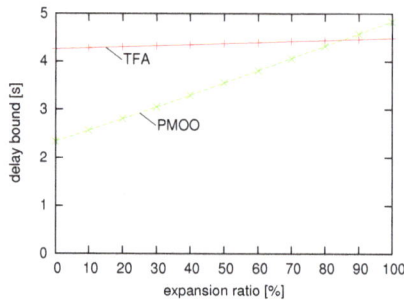

Figure 8. Investigation of PMOO vs. TFA under the drifting apart of maximum and minimum scaling curves.

Again, the PMOO performs superior to the TFA, and a certain amount of drifting between maximum and a minimum scaling curve are reached. After this point, the TFA becomes actually better than the PMOO as the uncertainty about the scaling becomes too large. Therefore, it can be wise to perform both SensorNC analyses.

5. Towards Self-Modeling Sensor Networks

In large-scale WSNs, any form of centralized control is often undesired if not impossible. Instead, a self-organized paradigm is followed; thus, it is consequent to also perform the modeling in a distributed, network-internal manner. The model is supposed to allow the WSN to act accordingly, e.g., by distributed decision on the admission of certain additional sensing tasks.

For such a self-modeling with SensorNC, the two basic pieces of modeling information are: the residual service a sensor can offer, i.e., the left-over service curve, and the arrival curve valid for a flow at a certain location in the network. The left-over service curve is the result of an analysis that requires the bound on interfering traffic. Deriving the arrival curves for interfering traffic, on the other hand, requires executing the deconvolution of flow arrivals with (left-over) service curves according to the exact shape of the sink tree. For this reason, general network calculus relies on the global view of the network and thus does not support self-modeling in WSNs. In this section, we present a way how, in a restricted, but practically very relevant setting, the SensorNC computations can be distributed over the WSN's sink tree network such that self-modeling is realized. We achieve self-modeling by rephrasing the SensorNC analysis, in particular the output arrival curve presented in Sections 3.2 and 3.3, as well as the PMOO analysis in Algorithm 2 where these curves are used. In order to do so, we need to restrict the curve shapes for modeling and analysis to the oftentimes used rate-latency service curves $\beta = \beta_{R,L} \in \mathcal{F}_{RL}$ and token-bucket arrival curves $\alpha = \gamma_{r,b} \in \mathcal{F}_{TB}$. Then, the three network calculus bounds can be computed as follows:

Corollary 3. *(Performance bounds in SensorNC) Consider a sensor node that offers a service curve $\beta = \beta_{R,T} \in \mathcal{F}_{RL}$. Assume a flow f with arrival curve $\alpha = \gamma_{r,b} \in \mathcal{F}_{TB}$ traversing this node. Then, we obtain the following SensorNC performance bounds:*

$$\text{output arrival curve: } \gamma_{r',b'}(d) = (\gamma_{r,b} \oslash \beta_{R,T})(d) = \begin{cases} 0 & \text{if } d = 0 \\ \gamma_{r,b+r\cdot T}(d) & \text{otherwise,} \end{cases}$$

$$\text{backlog: } \forall t \in \mathbb{R}^+ : B(t) = b + r \cdot T,$$

$$\text{delay: } \forall t \in \mathbb{R}^+ : D(t) = T + \frac{b}{R}.$$

Calculating the bound on network-internal traffic at any location uses aggregation and deconvolution. Given the above restriction of curve shapes, the following property holds for the combination of these operations.

Lemma 1. *(Distributivity of $\oslash$ with respect to +) For any $\alpha^{f_1}, \alpha^{f_2} \in \mathcal{F}_{TB}$ and $\beta \in \mathcal{F}_{RL}$, it holds that:*

$$\left(\alpha^{f_1} + \alpha^{f_2}\right) \oslash \beta = \alpha^{f_1} \oslash \beta + \alpha^{f_2} \oslash \beta.$$

Proof. Let $\alpha^{f_1} = \gamma_{r_1,b_1}$, $\alpha^{f_2} = \gamma_{r_2,b_2}$ and $\beta = \beta_{R,T}$. From Corollary 3, it follows that:

$$\left(\alpha^{f_1} + \alpha^{f_2}\right) \oslash \beta = \left((\gamma_{r_1,b_1} + \gamma_{r_2,b_2}) \oslash \beta_{R,T}\right)(d)$$
$$= \left(\gamma_{r_1+r_2, b_1+b_2} \oslash \beta_{R,T}\right)(d).$$

If $d = 0$, we have $\alpha^{f_1} \oslash \alpha^{f_2}(d) = 0$, and for $d > 0$, we get:

$$\left(\gamma_{r_1+r_2, b_1+b_2} \oslash \beta_{R,T}\right)(d) = \left(\gamma_{r_1+r_2, (b_1+b_2)+(r_1+r_2)\cdot T}\right)(d)$$
$$= \left(\gamma_{r_1, b_1+r_1\cdot T} + \gamma_{r_2, b_2+r_2\cdot T}\right)(d)$$
$$= \left(\gamma_{r_1, b_1+r_1\cdot T}\right)(d) + \left(\gamma_{r_2, b_2+r_2\cdot T}\right)(d)$$
$$= \left(\gamma_{r_1,b_1} \oslash \beta_{R,T}\right)(d) + \left(\gamma_{r_2,b_2} \oslash \beta_{R,T}\right)(d)$$
$$= \left(\alpha^{f_1} \oslash \beta\right)(d) + \left(\alpha^{f_2} \oslash \beta\right)(d).$$

□

Moreover, the composition rule of $\oslash$ follows from $f \oslash g \oslash h = f \oslash (g \otimes h)$ [6] by an argumentation similar to Lemma 1. The above distributivity and the composition rule allow one to compute the network internal output arrival curve of Algorithm 2 with a new SensorNC concatenation theorem.

Theorem 7. *(SensorNC concatenation theorem) Consider a sensor node s_i in a sink tree that is crossed by a set of flows $F = f_1, \ldots, f_n$ with arrival curves $\alpha^{f_1}, \ldots, \alpha^{f_n} \in \mathcal{F}_{TB}$. For the purpose of their aggregate output arrival curve calculation, the share of service offered to each flow $f \in F$ from its source to s_i within this aggregate is the concatenation of the service on its path. Then, the entire flow aggregate's output is bounded by*

$$\overline{\alpha}_i = \sum_{f \in F} \left(\alpha^f \oslash \bigotimes_{j=0}^{L(f,s_i)} \beta_{P(f,j)} \right)$$

where $L(f, s)$ is the location (index) of sensor node s on f's path, and in reverse, $P(f, i)$ denotes the sensor at location (index) i on f's path. Applying Corollary 3 and the composition rule, we can rephrase the equation to:

$$\overline{\alpha}_i = \gamma_{\overline{r}_i, \overline{b}_i}$$

with

$$\overline{r}_i = \sum_{f \in F} r^f \quad and \quad \overline{b}_i = \sum_{f \in F} \left(b^f + r^f \cdot \sum_{j=1}^{L(f,s_i)} T_{P(f,j)} \right).$$

Proof. Can be found in [22]. $\square$

Using this result in the PMOO analysis of Algorithm 2 requires separating the flow of interest from its interfering traffic. To do so, Theorem 7 can be applied to a the subset of flows $F' \subseteq F$ not including the flow of interest.

This reformulated computation's decisive improvement for a self-modeling WSN is the change from a total output arrival curve for an flow aggregate to an aggregate of flow-local results. The computations only make use of individual flow's arrival curves and the service curves of servers they cross. Information about the sink tree network's exact shape, merging locations with other flows or left-over service curves, are not required. Figure 9 illustrates the flow-locality of the computation.

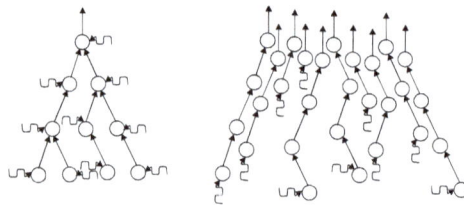

Figure 9. Sink tree network with flow aggregation (left) converted to a set of flow-local views by Theorem 7 (right).

Applying the new method from Theorem 7 instead of the conventional one of Section 3.3 has several practical advantages for self-modeling WSNs:

Complete model with low communication overhead: Flow-locality enables one to overcome the need for an additional protocol such as Deluge [23], distributing the information required to derive network-internal arrival curves and left-over service curves. Each flow's arrival at a server can be calculated hop-by-hop without compromising accuracy. A flow can carry information about its current arrival bound as payload, pushing the information to all sensors concerned. From all of the flows crossing them, sensor nodes can derive their local delay bound and backlog bound, as well as their left-over service curves. Thus, the network model is updated continuously, i.e., independent of a polling interval, with only a small payload overhead.

Local reaction to changes: The flow-locality also affects the recomputation effort in the case of parameter modifications. Using the conventional method, the locality of a modified parameter did not matter much due to the complex setting of flow aggregation locations; a change to a single parameter always invalidated a large amount of the derivation's intermediate results and triggered expensive recomputations, usually of the entire subtree. Theorem 7 prevents such an invalidation from spreading to flows not directly affected by a change, e.g., flows not crossing a sensor that adapted its rate, and thus enables a quick reaction to changes.

6. Tool Support

The DISCO Deterministic Network Calculator (DiscoDNC) [20,21] offers tool support for network calculus in general, as well as specialized SensorNC optimizations. For both, the DiscoDNC builds upon an implementation of piecewise linear curves and the network calculus operations. A recent overview of its core architecture can be found in [24].

The DiscoDNC implements the three analysis methods on the path of the flow of interest discussed in Section 3.4, TFA, SFA and PMOO analysis, as well as some more [25,26]. Additionally, the DiscoDNC provides methods to bound interfering flows' descriptions when they enter the flow of interest's path. Alternatives either implement the pay bursts only once principle (Section 2) or the PMOO principle. They either aggregate the interfering flows or execute a per-flow computation during the computation [27], and all alternatives can be augmented with an additional output burstiness restriction [28]. Thus, the DiscoDNC comprehensively covers a wide range of alternative analyses proceeding from generic network calculus.

For SensorNC, only a subset of these alternatives is relevant. For example, neither TFA nor SFA can outperform PMOO on the flow of interest's path. On the other hand, SensorNC offers specific solutions exploiting the oftentimes used token-bucket arrival curves and rate-latency service curves and the network's sink tree topology. For both of these specialized settings, the DiscoDNC implements separate code paths, as well. In particular, efficient computations of convolution and deconvolution of token-bucket arrival curves and rate-latency service curves are provided alongside the SensorNC concatenation theorem presented in Section 5. These specialized code paths are also suitable to demonstrate the reduced computational demand of SensorNC compared to generic network calculus [22]. Figure 10 shows such a comparison:

In Figure 10a, every flow's delay bound in a homogeneous sink tree with 100 sensor nodes is computed. Arrival curves are set to $\alpha = \gamma_{1,1}$; service curves are $\beta = \beta_{75,1}$; and the sink tree is randomly created with a maximum of five child nodes per sensor, as well as a maximum sink tree depth of 20. Whereas an alternative approach to achieve flow locality for self-modeling results in considerably worse delay bounds, the generic network calculus approach without flow locality and the SensorNC approach presented in the previous section yield the same delay bounds.

For a second set of results, we randomly created 40 sink trees of sizes up to 1000 sensor nodes. Each flow in each sink tree network was analyzed with generic network calculus (no flow locality) and SensorNC. The reduction of computation times achieved by SensorNC is shown in Figure 10b. Whereas the reduction is negligible in very small sink trees, it first rapidly increases and stays at a level of about five-times, although slowly decreasing with the WSN's size. Thus, the available tool support provided by the DiscoDNC shows that SensorNC can achieve the same accuracy as generic network calculus, yet at a fraction of the computational cost; a crucial improvement for deployment in self-modeling WSNs.

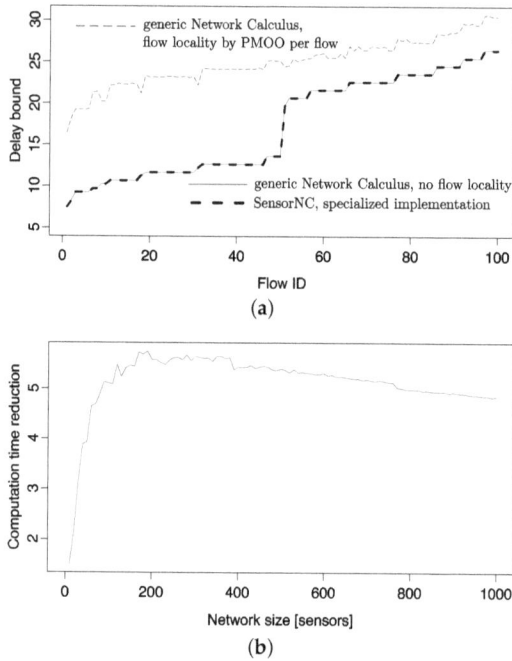

Figure 10. Generic network calculus computations vs. sensor network calculus (SensorNC): (**a**) delay bounds and (**b**) computation time reduction.

7. SensorNC Applications

In this section, we display several application examples where the SensorNC framework was instrumental.

7.1. Optimal Sink Placement

In many WSN applications, it is desired to collect the information acquired by sensors for processing, archiving and other purposes. The station where that information is required is usually the sink. For large-scale WSNs, a single-sink model is not scalable since message transfer delays, as well as energy consumption of the sensor nodes become prohibitive, due to the fact that most of the nodes would be far away from the sink, and thus, many hops must be traversed before the sink is reached. As a result, response times become excessive, and the lifetime of the WSN becomes very short. Therefore, it is sensible to deploy multiple sinks so that messages reach their destination with less hops, and consequently, response times are decreased and energy saved.

If sinks are placed in good locations, this can reduce traffic flow and energy consumption for sensor nodes. In particular, we focus on strategies to minimize the maximum worst-case delay, which is important for any timely actuation based on the information collected by a WSN.

In mathematical terms, we can pose the problem as follows:

$$min. \max_{i \in \{1,\ldots,n\}} \{D_i\}$$

with:

$$D_i = f\left(\tau | \vec{\alpha}, \vec{\beta}\right)$$

$$\tau = g\left(\vec{s}\mid\vec{p}, \mathcal{R}\right)$$

$$\vec{p} = \left(p_i^{(x)}, p_i^{(y)}\right)_{i=1,\dots,n}$$

$$\vec{s} = \left(s_j^{(x)}, s_j^{(y)}\right)_{j=1,\dots,k}$$

$$p_i, s_i \in \mathcal{F}$$

Here, n denotes the number of sensor nodes, and $\vec{p}$ is the vector of their locations in the sensor field $\mathcal{F}$; these locations are assumed to be given. The values D_i are the worst-case delays for each sensor node i. By minimizing the maximum worst-case delay in the field, it is ensured that response times are balanced as much as possible. k is the number of sinks, and the vector $\vec{s}$ contains their locations; these locations are the actual decision variables of the optimization problem. This is somewhat hidden by the fact that the delays D_i are only indirectly, affected by the choice of the sink locations, because as a first-order effect, the D_i are a function of the topology τ in which the WSN organized itself in the data flow towards the sinks and the arrival and service curves of the sensor nodes, denoted as $\vec{\alpha}$ and $\vec{\beta}$ in the formulas above. While the arrival and service curves are parameters for a given WSN scenario, the topology is, in turn, a function of the nodes' locations, the routing algorithm $\mathcal{R}$ and the sinks' locations $\vec{s}$, where however only the sinks' locations are variable, and the other two are again given parameters. Note, in particular, that we assume the routing algorithm to be given and not to be subject to the optimization. Although this could in principle be done, it would aggravate the problem further and is therefore left for further study.

Therefore, in principle, we face a continuous optimization problem where the objective function is to minimize the maximum worst-case delay in the field subject to constraints that ensure that each sensor node is connected to a sink (possibly via multi-hop communication, determined by the routing algorithm), as well as some geographic constraints. Due to the highly non-linear, jumpy behavior of the worst-case delay function f and thus of the objective function, which results from the formulas derived by sensor network calculus, the direct solution of that optimization problem is practically infeasible.

Therefore, a heuristic strategy for the sink placement problem that minimizes the maximum worst-case delay in WSNs based on the sensor network calculus framework was developed: genetic algorithm-based heuristic for sink placement (GASP). The crucial insight used by the GASP is that the continuous optimization can be reduced to a combinatorial optimization by identifying so-called regions of indifference, in which any location results in the same delay performance, and thus, a single location for such a region can be selected as a candidate. More details can be found in [29,30].

The performance of the GASP was analyzed in comparison to two other strategies. One is an optimal strategy, called OSP , based on an exhaustive search over the set of candidate locations (only feasible in small scenarios) which serves as an upper bound for the performance achievable by the GASP, and the other one is a Monte Carlo-based strategy, called MCP , that should serve as lower bound on the performance that can be expected from the GASP.

7.1.1. Small-Scale WSNs: Comparison between OSP and GASP

In this set of experiments, we analyze the worst-case delay for GASP and OSP for different, but relatively small network sizes of 30, 50 and 100 nodes. The number of candidate locations for the sinks were about 450, 950 and 2200 for the 30-, 50- and 100-node network, respectively. The number of sinks was restricted to two, since for the 100-node network, this already meant that the OSP strategy had to evaluate $\binom{2200}{2} = 2{,}418{,}900$ different combinations of sink placements. Each evaluation consists of 100 per-flow delay analyses, resulting in a total run-time of the OSP of several days. For the GASP, on the other hand, we chose a population size of 40 individuals, and the number of generations was set to 200, resulting in only 8000 different sink placements that where evaluated in only a few minutes. The worst-case delay results for the best sink placements found by the GASP and OSP are shown in Figure 11.

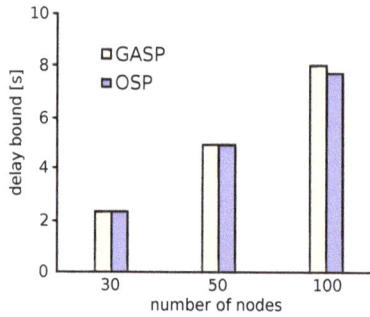

Figure 11. The worst case delay comparison of OSP vs. GASP.

As can be observed, the GASP performs very well in comparison to the OSP: for the 30- and 50-node network, it actually finds the global optimum; only for the 100-node case, the best sink placement found by the GASP lies slightly above the one found by the OSP (8.02 s vs. 7.70 s). That should be considered a success of the GASP since with a computational effort that is several orders of magnitude lower than for the OSP, it achieves almost the same quality of solutions. Whether this holds true for larger-scale scenarios is difficult to assess as the OSP is prohibitively computationally expensive. Therefore, in the next subsection, we compare the GASP against the MCP in larger-scale networks.

7.1.2. Large-Scale WSNs: Comparison between MCP and GASP

The question we address in this experiment is whether the GASP constitutes a more intelligent search strategy than a pure random search like the MCP. In fact, there would even be the possibility that the MCP could outperform the GASP. This would be the case if the GA operators were poorly designed and would mislead the GASP in areas of the search space that are fruitless. Therefore, in a certain sense, the following experiments also validate the design of the GASP heuristic.

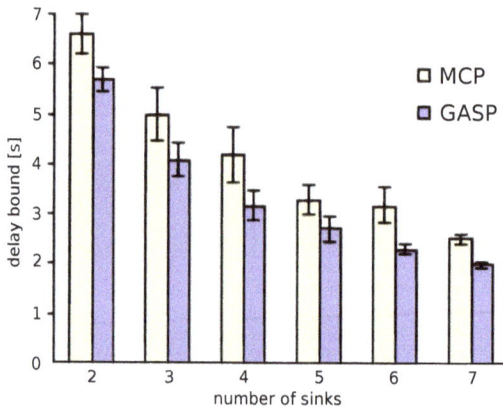

Figure 12. The worst case delay comparison of MCP vs. genetic algorithm-based heuristic for sink placement (GASP).

We investigate a 500-node network with up to seven sinks for 10 different scenarios, i.e., 10 different node distributions. For each of the scenarios, this resulted in approximately 13,000 candidate sink locations, so that at maximum, the search space becomes as big as $\binom{13,000}{7} \approx$

1.24×10^{25}. On the other hand, for the GASP, we used a population size of 100 with 100 generations until termination, resulting in 10,000 sink placements being evaluated for their worst-case delay. We allow the same amount of evaluations to the MCP. In any case, it is clear that this amount of sink placement evaluations constitutes only a tiny fraction of the overall search space.

The results (averaged over the 10 scenarios) of these experiments are shown in Figure 12. This analysis shows that the GASP performs better than the MCP at the same amount of computational effort. For the GASP strategy, the worst-case delay improves from 12 s to 5.7 s to 4.1 s to 3.2 s to 2.7 s to 2.3 s to 2.0 s for the one- to seven-sink scenarios, respectively. For the MCP, the worst-case delay improves from 14.1 s to 6.6 s to 5.0 s to 4.2 s to 3.3 s to 3.2 s to 2.5 s for the one- to seven-sink scenarios, respectively. The delay difference between the two strategies is roughly between 0.5 s and 2 s. This shows that the GA operators do something sensible as the GA search improves on the pure random search of the MCP. Besides, these numbers also give a feeling what the delay vs. number of sinks tradeoff looks like. Moreover, the confidence interval of MCP varies from 0.1 s to 1.2 s, whereas GASP varies from 0.05 s to 0.8 s. Obviously, providing a second sink improves the delay performance very much, whereas further sinks have a lesser effect on the maximum worst-case delay observed.

7.2. Node Placement Strategies

Node placement is a fundamental issue to be solved in WSNs. A proper node placement scheme can reduce the complexity of problems in WSNs, for example routing, data fusion, communication, etc. Furthermore, it can extend the lifetime of WSNs by minimizing energy consumption. In this section, we investigate the worst-case delay using SensorNC in random and deterministic node placements for large-scale WSNs under the following performance metrics. A more comprehensive investigation can be found in [31]. We consider three competitors: a uniform random, a square grid and a pattern-based tri-hexagon tiling (THT) node placement. We assume homogeneous sensor nodes.

7.2.1. Node Placement Schemes

During the design phase of WSNs, the designer only knows the number of sensor nodes, n. A circular field with radius R is considered in our experiments. Next, we introduce three node placement schemes.

Uniform Random

We choose a uniform random placement as one of the competitors. In the uniform random placement, each of the n sensors has equal probability of being placed at any point inside a given field, as shown in Figure 13a. For example, such a placement can result from throwing sensor nodes from an airplane, helicopter or UAV

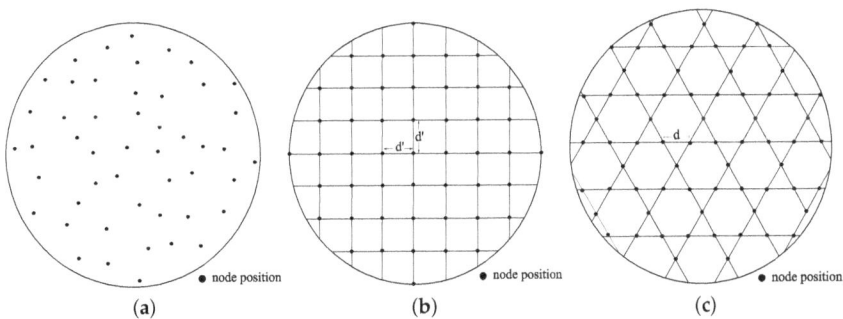

Figure 13. (a) Random, (b) square grid and (c) tri-hexagon tiling (THT) node placement.

Square Grid

Popular grid layouts are a unit square, an equilateral triangle, a hexagon, etc. Among them, we investigate the square grid because of its natural placement strategy over a unit square. Figure 13b shows a grid deployment of *n* sensors in a circular field.

Tri-Hexagon Tiling

The third strategy is based on tiling. A tiling is the covering of the entire plane with figures that do not overlap, nor leave any gaps. Among different tilings, we use a semi-regular tiling (which has exactly eight different tilings) where every vertex uses the same set of regular polygons. A regular polygon has the same side lengths and interior angles. We consider a semi-regular tiling that uses triangle and hexagon in the two-dimensional plane, the so-called 3-6-3-6 tri-hexagon tiling. The name comes from going around a vertex and listing the number of sides each regular polygon has, as illustrated in Figure 13c. Here, we combine the advantages of a triangle grid and a hexagon grid.

7.2.2. Results

The primary factors for all experiments are: the number of nodes, the number of sinks and the sensing range. For each deployment, nodes are distributed over a circular field shape, and sinks are placed at the center of gravity of a sector of a circle (CGSC). In the random deployment, we generated 10 scenarios and took the average value for the analysis of the worst-case delay in the sensor field. The routing topology we use here is based on Dijkstra's shortest path algorithm, which produces the shortest hop distance from a source to a sink. We also assume that r_{tx} is twice r_{sense} in all strategies. All of the selected values for the experiments are based on a realistic model of an MICAz mote running under TinyOS.

The analytical results of the worst-case delay comparison among the three strategies are shown in Figure 14. For SensorNC computations, the token-bucket arrival curve and rate-latency service curves are considered. In particular, for the service curve, we use a rate-latency function, which corresponds to a duty cycle of 1%. For a 1% duty cycle, it takes 5 ms of time-on-duty with a 500-ms cycle length, which results in a latency of 0.495 s (the values are calculated based on CC2420AckLpl.h and CC2420AckLplP.nc.) The corresponding forwarding rate is 2500 bps. In all scenarios, THT outperforms the other strategies. In a 100-node network, the worst-case delay improves from 2.04 s to 1.52 s to 1.5 s for the 2-, 3- and 4-sink scenarios, respectively. In fact, the more sinks, the lower the worst-case delay should be. However, the sink placement at CGSC does not perform so well for a larger number of sinks. Another interesting observation is that the random deployment can have a lower worst-case delay than the square grid deployment, e.g., for a 1000-node network with 20 sinks scenario. It seems that a random deployment is more or less comparable to a square grid for a large-scale network.

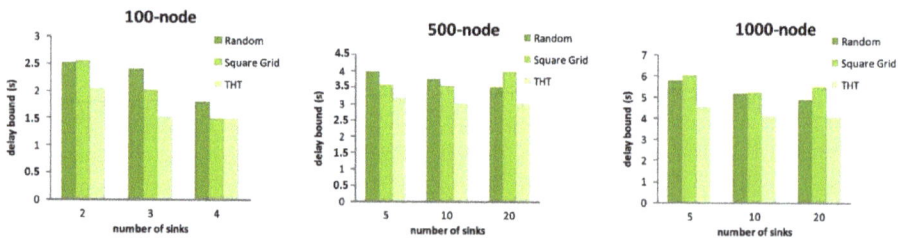

Figure 14. Worst-case delay comparison of three-node placement strategies in different scenarios.

7.3. TDMA Optimization

In this application (see also [8]), we present a means to find energy-efficient medium assignments in time-slotted WSNs that satisfy given real-time constraints. Specifically, we present a way to find the optimal length of time slots and cycles in TDMA schemes. This problem is solved analytically to find solutions for a general sink tree network under a fluid SensorNC model.

When designing a TDMA system, a choice has to be made for how long the repetitive TDMA frame, as well as the individual slot sizes of each participating node are. For some network nodes, that just requires a short slot in which they can send collected data and perhaps receive an acknowledgment from a downstream node. However, in larger WSNs, some nodes have higher bandwidth requirements for forwarding other nodes' data, while collecting and sending data themselves. Aside from avoiding contention, using TDMA also reduces energy consumption by making it possible for nodes to power down in periods without relevant traffic.

Since in WSNs, two main concerns are minimizing power consumption and meeting delay bounds, while transmission bandwidth requirements tend to be low, we want to maximize the frame length, giving the sensor nodes the opportunity to disable their radio transceivers or even go into deep sleep modes.

7.3.1. General TDMA Design Problem

From those requirements, we formulate the TDMA design problem as an optimization problem for a tree network with n nodes where from each node, a flow is originating:

$$
\begin{aligned}
\text{max.} \quad & Z = \min_{1 \leq i \leq n} \{f - s_i\} \\
\text{s.t.} \quad & \sum_{i=1}^{n} s_i \leq f && \text{(TDMA integrity)} \\
& \forall i : d_i(f, \vec{s}|r, b, C) \leq D && \text{(Delay)} \\
& \forall i : \frac{s_i}{f} \cdot C \geq F_i r && \text{(Rate)} \\
& \forall i : s_i \geq 0,\ f \geq 0 && \text{(Non-negativity)}
\end{aligned}
$$

Here, f is the cycle length of the repetitive TDMA frame, and s_i is the amount of time devoted to node i for sending (slot size of node i). These constitute the decision variables. For the parameters of the problem, we further have D as the maximum permissible delay that may be incurred by any flow in the sensor field, $d_i(f, \vec{s}|r, b, C) = h(\gamma_{r,b}, \beta_i^{\text{l.o.}})$ as the actual maximum delay incurred by flow i (which is computed based on the PMOO analysis described in Algorithm 2), F_i as the number flows carried by node i (including the flow originating at node i), C as the medium rate ("capacity") and r as the maximum sustained rate for any flow, as well as b as the maximum burst of a flow. Note that we assume the sensors to have identical arrival curves $\gamma_{r,b}$, as well as an identical delay requirement D, which for many practical situations will be no restriction and makes the further analysis more tractable.

The objective function reflects the fact that the minimum sleeping period over all nodes in the field should be maximized, thus achieving a maximum lifetime of the network. The TDMA integrity constraint captures the fact that all slot sizes together must fit into the TDMA frame. Obviously, the delay target should be met for all flows, which is captured by the delay constraints; also, all of the rate constraints must be met in order not to obtain infinite delay bounds for the flows. Of course, we also have non-negativity constraints for the decision variables.

Unfortunately, this general modeling of the TDMA design problem results in a very hard to solve non-linear programming problem. The non-linearity is exhibited in the objective function, as well as in the delay constraints. Hence, the only viable approach is to simplify the problem structure if a solution shall be found for larger instances of the TDMA design problem. There are two intuitive approaches towards relaxing the problem:

1. Equal slot sizing (ESS): the assignment may be made such that inside a fixed time slot length, each node can transmit enough data to meet all requirements.

2. Traffic-proportional slot sizing (TPSS): slots may be assigned such that each node only claims the resources necessary to fulfil its own duties, depending on the input bandwidth and forwarded data streams.

While the second relaxation approach may appear more efficient, it is also harder to set up. The first approach requires rather little information—the number of nodes and the bandwidth requirements of the node serving the highest number of flows—and the second method requires good knowledge of the topology, which may not always be at hand. Furthermore, in [8], it is shown that ESS is actually superior to TPSS, which is why we further focus on it.

Two variables need to be controlled under ESS: the overall frame length f and the individual slot length s. Obviously, with an increasing frame length, a node may sleep longer between transmission or reception phases, but delay is increased at the same time. For a given f in a network with n nodes, s is limited to values between an upper bound $\frac{f}{n}$ and a lower bound that is given by the minimum bandwidth requirements.

Next, we state the TDMA design problem under ESS, which as we show below is amenable to an analytical solution under TDMA service curves (over-)approximated by rate-latency service curves.

TDMA Design under Equal Slot Sizing

Under ESS, we now consider a common slot size s for all nodes. The TDMA design problem can then be formulated as:

$$\begin{aligned}
\text{max.} \quad & Z = f - s \\
\text{s.t.} \quad & s \le \frac{f}{n} \\
& d(f, s | r, b, C) = \max_{1 \le i \le n} d_i(f, s | r, b, C) \le D \\
& \frac{s}{f} \cdot C \ge F_{max} r \\
& f \ge 0
\end{aligned}$$

We obtain an optimization problem with a linear objective function, and only four constraints, a great simplification. Most of the reduction in complexity is due to the lower number of decision variables as, e.g., for the rate constraints that collapse into a single one. For the delay constraints, this is not as easily seen, but the reader may ascertain her/himself that there is always one flow that is the worst-case flow and whose delay constraint is dominating all of the other flows' delay constraints as they are all facing the same situation when considering the parameters. For example, in a fully-occupied n-ary tree, one can choose any leaf node and its corresponding flow as the flow whose delay constraint is the dominating one.

7.3.2. Analytical Solution in the Fluid Setting

In the following, we first discuss the ESS relaxation in a simple, yet illustrative example of a two-hop sensor network, as shown in Figure 15.

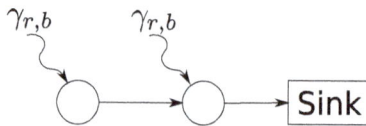

Figure 15. Two-node example network.

We assume the rate-latency curves as fluid approximations for the TDMA service in order to keep the problem analytically tractable. Based on this, we show how in the general sink tree case of fluid service curves we can derive an optimal solution.

Under ESS and in the two-hop network, we obtain the following incarnation of the optimal TDMA problem:

$$\max. \quad Z = f - s$$
$$\text{s.t.} \quad s \leq \frac{f}{2}$$
$$d_2(f, s|r, b, C) = \frac{\frac{s}{f}C(f-s)+2b}{\frac{s}{f}C-r} + f - s \leq D$$
$$\frac{s}{f} \cdot C \geq 2r$$
$$f \geq 0$$

Since the objective function is linear, the solution to this optimization problem must lie on the border of the feasible region (it is guaranteed to exist since the feasible region is closed). In Figure 16, the feasible region, as well as a contour line of the objective function are drawn (for $r = 1, b = 1, C = 10, D = 1$).

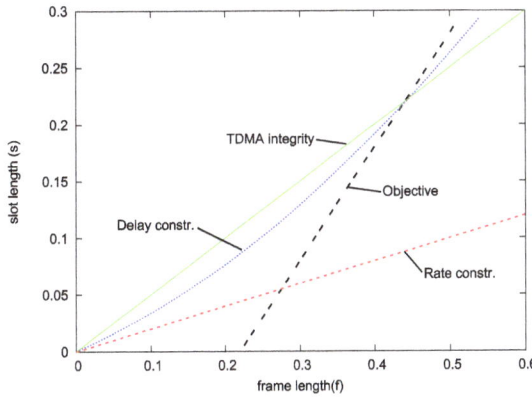

Figure 16. Graphical illustration of the optimization problem for Equal slot sizing (ESS). The feasible region is above the delay constraint and below the TDMA integrity constraint. The rate constraint does not affect the feasible region in this example.

It can be seen that the optimum must be taken on at the lower border of the feasible region. In fact, we can write the border constituted by the delay constraint as shown in Equation (6),

$$s = g(f|r, b, C, D) = \frac{\sqrt{C^2 D^2 + (6frC - 4fC^2)D + 4f^2 C^2 + (16bf - 4f^2 r)C + f^2 r^2 + 2fC - CD + fr}}{4C} \quad (6)$$

because the delay constraint is a quadratic form in s and f, which can be solved for s with two real solutions of which we take the larger one as it results in a more binding constraint. A moment's consideration exhibits that $\forall f : \frac{\partial g}{\partial f} < 1$, since otherwise, an increase in frame size would result in a larger increase of the slot size, which obviously cannot be the case. On the other hand, the partial derivative of the objective function after f is one, which means that the optimum must be taken on at the corner point of the feasible region where the delay constraint and TDMA integrity constraint intersect. In other words, the TDMA design problem under ESS can be reduced to matching the delay with the TDMA integrity constraint.

Analytical Solution for ESS in General Sink Trees

What remains to be done in general sink trees compared to the two-hop network in the previous setting is to show that the delay constraint again takes on a quadratic form. This then allows one to easily express the slot size as a function of the frame size, and the same arguments as in the two hop case lead to the conclusion that the optimum solution is given at the point where delay and TDMA integrity constraint are matched. Hence, let us discuss the delay constraint in a general sink tree network.

We assume a general sink tree network with each node offering a service curve $\beta_{\frac{s}{f}C, f-s}$ and flows starting from each node constrained by an arrival curve $\gamma_{r,b}$. Looking at a particular flow, we have a situation as depicted in Figure 17. Applying the PMOO analysis results in the following left-over service curve for the flow of interest:

$$
\begin{aligned}
\beta^{\text{l.o.}} &= \left[[\beta_{R,T} - \gamma_{r_1,b_1}]^+ \otimes \beta_{R,T} - \gamma_{r_2,b_2} \right]^+ \otimes \dots \\
&= \beta_{R - \sum_{i=1}^{n-1} r_i, \frac{T\left(nR - \sum_{i=1}^{n-1}\sum_{j=1}^{i} r_j\right) + \sum_{i=1}^{n-1} b_i}{R - \sum_{i=1}^{n-1} r_i}}
\end{aligned}
$$

with $R = \frac{s}{f}C$ and $T = f - s$ and $r_i = a_i r$ and $b_i = c_i b + d_i rT$. For the latter expressions, the parameters $a_i, c_i, d_i \in \mathbb{N}$ depend on the topology. The delay constraint for flow n (i.e., the one originating at node n) can thus be expressed as shown in Equation (7).

$$
\begin{aligned}
h\left(\gamma_{r,b}, \beta^n\right) &= \frac{b}{\frac{s}{f}C - r\sum_{i=1}^{n-1} a_i} \\
&+ \frac{(f-s)\left(n\frac{s}{f}C - r\sum_{i=0}^{n-1}\sum_{j=1}^{i} a_j\right) + b\sum_{i=1}^{n-1} c_i + r(f-s)\sum_{i=1}^{n-1} d_i}{\frac{s}{f}C - r\sum_{i=1}^{n-1} a_i} \qquad (7) \\
&\le D
\end{aligned}
$$

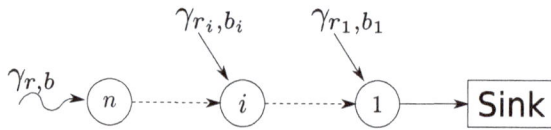

Figure 17. Flow in a general sink tree.

While seemingly a complex expression, this constitutes again a quadratic form in f and s, which can be recast to s, where we can ignore the smaller right-hand side version as only the larger one is physically meaningful. Hence, we can make the same observations as in the two-hop network case and again reason that the optimal solution must be taken on at the corner point of the feasible region where TDMA integrity and the delay constraint are matched.

7.4. Further Applications

There are further applications of SensorNC of which we briefly mention some here without going into further details:

An early application and customization of SensorNC was done in the context of ZigBee cluster tree networks [11,32]. Achieving (statistical) worst-case delay bounds despite uncertainty about the topology of a WSN has been addressed in [9,33]. Application of the SensorNC in the case of multiple sinks in the sensor field has been demonstrated in [34]. In [35], it was investigated how traffic splitting algorithms perform in terms of worst-case performance in mesh-based WSNs. A MAC protocol has been designed to exactly match the assumptions of SensorNC in order to enable an efficient and effective real-time communication behavior in WSNs [36]. An investigation of the power management in video sensor networks has been performed using SensorNC in [37]. The issue of topology control in combination with sink placement has been dealt with in [38] in the framework of SensorNC. Several efforts to compute energy-efficient trajectories for mobile sinks when delay guarantees also have to be met have been performed using SensorNC [39–41]. The effect of network coding on the QoS in a WSN has been investigated using SensorNC in [42]. Multimode WSNs have been analyzed using SensorNC in [43]. Industrial wireless mesh networks have been dimensioned using SensorNC in [44].

8. Discussion and Conclusions

In this article, we have presented the developments around the SensorNC framework since its inception in [4]. We have presented the basic mathematical methodology, as well as some of the most significant advancements such as the accommodation of in-network processing and self-modeling. From a practical point of view, we also emphasized the issue of tool support, which is by now well catered for by the DiscoDNC software. Applications of the SensorNC are abundant; a few have been discussed in some detail; several others have been mentioned.

As the SensorNC calculates performance bounds, an immediate question arises with respect to the accuracy of the bounds. This has been investigated via simulations and practical experiments, and the results speak in favor of SensorNC's precision [36,45]. Clearly, this also depends on many details of the respective WSN settings.

The work on the SensorNC should be continued. Clearly, some important challenges are still open. In particular, the wireless characteristics could be better reflected in a stochastic system model. There has been a huge amount of work along the lines of a general stochastic network calculus (see, e.g., [46] for a recent paper) and more specifically to model wireless channels, e.g., [47]. However, how to integrate these results into the SensorNC specifics is still open mainly due to the important issue of dealing with stochastic dependencies in larger topologies. Furthermore, the issue of retransmissions due to loss has been addressed in a single-node stochastic network calculus setting [48]. Furthermore, these results should be transferred suitably to the more challenging setting in SensorNC. Again, dealing with stochastic dependencies becomes a major mathematical challenge. Yet another advanced issue results from feedback control loops as, e.g., typical in cyber-physical systems [1]. Here, progress has been made recently in the stochastic setting, which could open up new opportunities for applications of the SensorNC [49–51].

At last, combinations of SensorNC with other formal verification techniques as in [52] could bring up new fruitful capabilities in the quest for WSNs with predictable performance.

Acknowledgments: Steffen Bondorf is supported by a Carl Zeiss Foundation grant.

Author Contributions: The original SensorNC framework was chiefly conceived by Jens Schmitt as well as the advanced aspects of in-network processing; Steffen Bondorf is the main contributor towards the self-modeling and tool aspects; Wint Yi Poe contributed substantially to the SensorNC applications. Jens Schmitt and Steffen Bondorf wrote the paper.

Conflicts of Interest: The authors declare no conflicts of interest.

References

1. Song, H.; Rawat, D.B.; Jeschke, S.; Brecher, C. *Cyber-Physical Systems: Foundations, Principles and Applications*; Morgan Kaufmann: Burlington, MA, USA, 2016.
2. Mao, X.; Miao, X.; He, Y.; Li, X.Y.; Liu, Y. CitySee: Urban CO_2 monitoring with sensors. In Proceedings of the IEEE Conference on Computer Communications (INFOCOM), Orlando, FL, USA, 25–30 March 2012; pp. 1611–1619.
3. Liu, Y.; Zhou, G.; Zhao, J.; Dai, G.; Li, X.Y.; Gu, M.; Ma, H.; Mo, L.; He, Y.; Wang, J.; et al. Long-term Large-scale Sensing in the Forest: Recent Advances and Future Directions of GreenOrbs. *Front. Comput. Sci. China* **2010**, *4*, 334–338.
4. Schmitt, J.B.; Roedig, U. Sensor Network Calculus—A Framework for Worst Case Analysis. In Proceedings of the First IEEE International Conference on Distributed Computing in Sensor Systems (DCOSS'05), Marina del Rey, CA, USA, 30 June–1 July 2005; pp. 141–154.

5. Chang, C.S. *Performance Guarantees in Communication Networks*; Springer: London, UK, 2000.

6. Le Boudec, J.Y.; Thiran, P. *Network Calculus: A Theory of Deterministic Queuing Systems for the Internet*; Springer: Berlin/Heidelberg, Germany, 2001.

7. Baccelli, F.; Cohen, G.; Olsder, G.J.; Quadrat, J.P. *Synchronization and Linearity: An Algebra for Discrete Event Systems*; John Wiley & Sons Ltd.: Hoboken, NJ, USA, 1992.

8. Gollan, N.; Schmitt, J.B. Energy-Efficient TDMA Design Under Real-Time Constraints in Wireless Sensor Networks. In Proceedings of the 15th International Symposium on Modeling, Analysis, and Simulation of Computer and Telecommunication Systems, 2007 (MASCOTS '07), Istanbul, Turkey, 24–26 October 2007; pp. 80–87.

9. Bondorf, S.; Schmitt, J.B. Statistical response time bounds in randomly deployed wireless sensor networks. In Proceedings of the 35th IEEE Local Computer Network Conference (LCN), Denver, CO, USA, 10–14 October 2010; pp. 340–343.

10. Cattelan, B.; Bondorf, S. Iterative Design Space Exploration for Networks Requiring Performance Guarantees. In Proceedings of the 36th IEEE/AIAA Digital Avionics Systems Conference (DASC 2017), St. Petersburg, FL, USA, 16–21 September 2017.

11. Koubâa, A.; Alves, M.; Tovar, E. Modeling and Worst-Case Dimensioning of Cluster-Tree Wireless Sensor Networks. In Proceedings of the 27th IEEE International Real-Time Systems Symposium, 2006 (RTSS '06), Rio de Janeiro, Brazil, 5–8 December 2006; pp. 412–421.

12. Bouillard, A.; Thierry, É. An Algorithmic Toolbox for Network Calculus. *Discret. Event Dyn. Syst.* **2008**, *18*, 3–49.

13. Lampka, K.; Bondorf, S.; Schmitt, J.B.; Guan, N.; Yi, W. Generalized Finitary Real-Time Calculus. In Proceedings of the 36th IEEE International Conference on Computer Communications (INFOCOM 2017), Atlanta, GA, USA, 1–4 May 2017.

14. Schmitt, J.B.; Zdarsky, F.A.; Martinovic, I. Improving Performance Bounds in Feed-Forward Networks by Paying Multiplexing Only Once. In Proceedings of the GI/ITG International Conference on Measurement, Modelling and Evaluation of Computer and Communication Systems (MMB), Dortmund, Germany, 31 March–2 April 2008; pp. 1–15.

15. Schmitt, J.; Zdarsky, F.A.; Fidler, M. Delay Bounds under Arbitrary Multiplexing: When Network Calculus Leaves You in the Lurch. In Proceedings of the 27th IEEE International Conference on Computer Communications (INFOCOM 2008), Phoenix, AZ, USA, 13–18 April 2008.

16. Wandeler, E.; Maxiaguine, A.; Thiele, L. Quantitative Characterization of Event Streams in Analysis of Hard Real-Time Applications. In Proceedings of the 10th IEEE Real-Time and Embedded Technology and Applications Symposium (RTAS 2004), Toronto, ON, Canada, 25–28 May 2004; pp. 450–461.

17. Fidler, M.; Schmitt, J. On the Way to a Distributed Systems Calculus: An End-to-End Network Calculus with Data Scaling. In Proceedings of the Joint International Conference on Measurement and Modeling of Computer Systems, SIGMETRICS/Performance 2006, Saint Malo, France, 26–30 June 2006; pp. 287–298.

18. Schmitt, J.B.; Zdarsky, F.A.; Thiele, L. A Comprehensive Worst-Case Calculus for Wireless Sensor Networks with In-Network Processing. In Proceedings of the 28th IEEE International Real-Time Systems Symposium (RTSS 2007), Tucson, AZ, USA, 3–6 December 2007; pp. 193–202.

19. Crossbow Technology Inc. *Mote Processor Radio (MPR) Platforms and Mote Interface Boards (MIB), Review B*; Crossbow Technology Inc.: San Jose, CA, USA, 2006.

20. Schmitt, J.B.; Zdarsky, F.A. The DISCO Network Calculator—A Toolbox for Worst Case Analysis. In Proceedings of the 1st International Conference on Performance Evaluation Methodolgies and Tools (ValueTools '06), Pisa, Italy, 11–13 October 2006.

21. Bondorf, S.; Schmitt, J.B. The DiscoDNC v2—A Comprehensive Tool for Deterministic Network Calculus. In Proceedings of the 8th International Conference on Performance Evaluation Methodologies and Tools (ValueTools), Bratislava, Slovakia, 9–11 December 2014.

22. Bondorf, S.; Schmitt, J.B. Boosting sensor network calculus by thoroughly bounding cross-traffic. In Proceedings of the IEEE Conference on Computer Communications (INFOCOM), Hong Kong, China, 26 April–1 May 2015; pp. 235–243.

23. Hui, J.W.; Culler, D. The Dynamic Behavior of a Data Dissemination Protocol for Network Programming at Scale. In Proceedings of the 2nd International Conference on Embedded Networked Sensor Systems (SenSys '04), Baltimore, MD, USA, 3–5 November 2004; pp. 81–94.

24. Schon, P.; Bondorf, S. Towards Unified Tool Support for Real-time Calculus & Deterministic Network Calculus. In Proceedings of the 29th Euromicro Conference on Real-Time Systems (ECRTS 2017), Dubrovnik, Croatia, 28–30 June 2017.

25. Bondorf, S.; Nikolaus, P.; Schmitt, J.B. Quality and Cost of Deterministic Network Calculus—Design and Evaluation of an Accurate and Fast Analysis. In Proceedings of the ACM SIGMETRICS International Conference on Measurement and Modeling of Computer Systems (SIGMETRICS 2017), Urbana-Champaign, IL, USA, 5–9 June 2017.

26. Bondorf, S. Better Bounds by Worse Assumptions—Improving Network Calculus Accuracy by Adding Pessimism to the Network Model. In Proceedings of the IEEE International Conference on Communications (ICC 2017), Paris, France, 21–25 May 2017.

27. Bondorf, S.; Schmitt, J.B. Calculating Accurate End-to-End Delay Bounds—You Better Know Your Cross-Traffic. In Proceedings of the 9th International Conference on Performance Evaluation Methodologies and Tools (ValueTools), Berlin, Germany, 14–16 December 2015; pp. 17–24.

28. Bondorf, S.; Schmitt, J.B. Improving Cross-Traffic Bounds in Feed-Forward Networks—There is a Job for Everyone. In Proceedings of the GI/ITG International Conference on Measurement, Modelling and Evaluation of Dependable Computer and Communication Systems (MMB & DFT), Münster, Germany, 4–6 April 2016.

29. Poe, W.Y.; Schmitt, J. *Minimizing the Maximum Delay in Wireless Sensor Networks by Intelligent Sink Placement*; Technical Report 362/07; University of Kaiserslautern: Kaiserslautern, Germany, 2007.

30. Poe, W.Y.; Schmitt, J. Placing Multiple Sinks in Time-Sensitive Wireless Sensor Networks Using a Genetic Algorithm. In Proceedings of the 14th GI/ITG Conference on Measurement, Modeling, and Evaluation of Computer and Communication Systems (MMB 2008), Dortmund, Germany, 31 March–2 April 2008.

31. Poe, W.Y.; Schmitt, J. Node Deployment in Large Wireless Sensor Networks:Coverage, Energy Consumption, and Worst-Case Delay. In Proceedings of the 5th ACM SIGCOMM Asian Internet Engineering Conference (AINTEC), Bangkok, Thailand, 18–20 November 2009.

32. Jurcik, P.; Koubâa, A.; Severino, R.; Alves, M.; Tovar, E. Dimensioning and Worst-Case Analysis of Cluster-Tree Sensor Networks. *ACM Trans. Sens. Netw.* **2010**, *7*, 1–47.

33. Schmitt, J.B.; Roedig, U. Worst Case Dimensioning of Wireless Sensor Networks under Uncertain Topologies. In Proceedings of the Workshop on Resource Allocation in Wireless Networks at IEEE WiOpt, Trento, Italy, April 2005.

34. Schmitt, J.B.; Zdarsky, F.A.; Roedig, U. Sensor Network Calculus with Multiple Sinks. In Proceedings of the Performance Control in Wireless Sensor Networks Workshop at the IFIP Networking, Coimbra, Portugal, 15–19 May 2006; pp. 6–13.

35. She, H.; Lu, Z.; Jantsch, A.; Zhou, D.; Zheng, L.R. Performance Analysis of Flow-Based Traffic Splitting Strategy on Cluster-Mesh Sensor Networks. *Int. J. Distrib. Sens. Netw.* **2012**, doi:10.1155/2012/232937.

36. Suriyachai, P.; Roedig, U.; Scott, A.C.; Gollan, N.; Schmitt, J.B. Dimensioning of Time-Critical WSNs—Theory, Implementation and Evaluation *JCM* **2011**, *6*, 360–369.

37. Cao, Y.; Xue, Y.; Cui, Y. Network-Calculus-Based Analysis of Power Management in Video Sensor Networks. In Proceedings of the Global Communications Conference, 2007 (GLOBECOM '07), Washington, DC, USA, 26–30 November 2007; pp. 981–985.

38. Safa, H.; El-Hajj, W.; Zoubian, H. A Robust Topology Control Solution for the Sink Placement Problem in WSNs. *J. Netw. Comput. Appl.* **2014**, *39*, 70–82.

39. Jurcík, P.; Severino, R.; Koubâa, A.; Alves, M.; Tovar, E. Real-Time Communications Over Cluster-Tree Sensor Networks with Mobile Sink Behaviour. In Proceedings of the 2008 14th IEEE International Conference on Embedded and Real-Time Computing Systems and Applications, Kaohsiung, Taiwan, 25–27 August 2008; pp. 401–412.

40. Poe, W.Y.; Beck, M.A.; Schmitt, J. Planning the Trajectories of Multiple Mobile Sinks in Large-Scale, Time-Sensitive WSNs. In Proceedings of the 7th IEEE International Conference on Distributed Computing in Sensor Systems (DCOSS 2011), Barcelona, Spain, 27–29 June 2011.

41. Poe, W.Y.; Beck, M.A.; Schmitt, J.B. Achieving High Lifetime and Low Delay in Very Large Sensor Networks using Mobile Sinks. In Proceedings of the 8th IEEE International Conference on Distributed Computing in Sensor Systems (DCOSS), Hangzhou, China, 16–18 May 2012; pp. 17–24.

42. Jiang, L.; Yu, L.; Chen, Z. Network calculus based QoS analysis of network coding in Cluster-tree wireless sensor network. In Proceedings of the 2012 Computing, Communications and Applications Conference, Hong Kong, China, 11–13 January 2012; pp. 1–6.

43. Jin, X.; Guan, N.; Wang, J.; Zeng, P. Analyzing Multimode Wireless Sensor Networks Using the Network Calculus. *J. Sens.* **2015**, *2015*, 851608.

44. Shang, Z.J.; Cui, S.J.; Wang, Q.S. Network Calculus Based Dimensioning for Industrial Wireless Mesh Networks. *Appl. Mech. Mater.* **2013**, *303–306*, 1989–1995.

45. Roedig, U.; Gollan, N.; Schmitt, J.B. Validating the Sensor Network Calculus by Simulations. In Proceedings of the Performance Control in Wireless Sensor Networks Workshop (WICON '07), Austin, TX, USA, 22–24 October 2007.

46. Ciucu, F.; Schmitt, J. Perspectives on Network Calculus—No Free Lunch, But Still Good Value. In Proceedings of the ACM SIGCOMM 2012 Conference on Applications, Technologies, Architectures, and Protocols for Computer Communications (SIGCOMM '12), Helsinki, Finland, 13–17 August 2012; ACM: New York, NY, USA, 2012; pp. 311–322.

47. Schiessl, S.; Naghibi, F.; Al-Zubaidy, H.; Fidler, M.; Gross, J. On the delay performance of interference channels. In Proceedings of the 2016 IFIP Networking Conference, Networking 2016 and Workshops, Vienna, Austria, 17–19 May 2016; pp. 216–224.

48. Wang, H.; Schmitt, J.; Ciucu, F. Performance Modelling and Analysis of Unreliable Links with Retransmissions using Network Calculus. In Proceedings of the 25th International Teletraffic Congress (ITC 25), Shanghai, China, 10–12 September 2013.

49. Beck, M.A.; Schmitt, J.B. Window Flow Control in Stochastic Network Calculus—The General Service Case. In Proceedings of the 9th International Conference on Performance Evaluation Methodologies and Tools (Valuetools), Berlin, Germany, 14–16 December 2015; pp. 25–32.

50. Beck, M.A.; Schmitt, J.B. Generalizing Window Flow Control in Bivariate Network Calculus to Enable Leftover Service in the Loop. *Perform. Eval.* **2016**, *114*, 45–55.

51. Shekaramiz, A.; Liebeherr, J.; Burchard, A. Network Calculus Analysis of a Feedback System with Random Service. In Proceedings of the 2016 ACM SIGMETRICS International Conference on Measurement and Modeling of Computer Science, Antibes Juan-Les-Pins, France, 14–18 June 2016; pp. 393–394.

52. Mouradian, A.; Blum, I.A. Formal Verification of Real-Time Wireless Sensor Networks Protocols: Scaling Up. In Proceedings of the 26th Euromicro Conference on Real-Time Systems, Madrid, Spain, 8–11 July 2014; pp. 41–50.

Journal of
Sensor and
Actuator Networks

MDPI

Article

Athena: Towards Decision-Centric Anticipatory Sensor Information Delivery [†]

Jongdeog Lee [1], Kelvin Marcus [2], Tarek Abdelzaher [1,*], Md Tanvir A. Amin [1], Amotz Bar-Noy [3], William Dron [4], Ramesh Govindan [5], Reginald Hobbs [2], Shaohan Hu [1], Jung-Eun Kim [1], Lui Sha [1], Shuochao Yao [1] and Yiran Zhao [1]

[1] University of Illinois at Urbana-Champaign, Urbana, IL 61801, USA; jlee700@illinois.edu (J.L.); tanviralamin@gmail.com (M.T.A.A.); shu17@illinois.edu (S.H.); jekim314@illinois.edu (J.-E.K.); lrs@illinois.edu (L.S.); yao9@illinois.edu (S.Y.); zhao97@illinois.edu (Y.Z.)
[2] U.S. Army Research Laboratory, Adelphi, MD 20783, USA; kelvin.m.marcus.civ@mail.mil (K.M.); reginald.l.hobbs2.civ@mail.mil (R.H.)
[3] The City University of New York, New York, NY 10016, USA; amotz@sci.brooklyn.cuny.edu
[4] Raytheon BBN Technologies, Cambridge, MA 02138, USA; wdron@bbn.com
[5] University of Southern California, Los Angeles, CA 90089, USA; ramesh@usc.edu
[*] Correspondence: zaher@illinois.edu
[†] This paper is an extended version of Abdelzaher, T.; Amin, T.A.; Bar-Noy, A.; Dron, W.; Govindan, R.; Hobbs, R.; Hu, S.; Kim, J.-E.; Yao, S.; Zhao, Y. Decision-driven Execution: A Distributed Resource Management Paradigm for the Age of IoT. In Proceedings 37th IEEE International Conference on Distributed Computing Systems (ICDCS), Atlanta, GA, USA, 5–8 June 2017; IEEE: Piscataway, NJ, USA, 2017.

Received: 4 September 2017; Accepted: 8 January 2018; Published: 15 January 2018

Abstract: The paper introduces a new direction in quality-of-service-aware networked sensing that designs communication protocols and scheduling policies for data delivery that are optimized specifically for decision needs. The work complements present decision monitoring and support tools and falls in the larger framework of decision-driven resource management. A hallmark of the new protocols is that they are aware of the inference structure used to arrive at decisions (from logical predicates), as well as the data (and data quality) that need to be furnished to successfully evaluate the unknowns on which these decisions are based. Such protocols can therefore anticipate and deliver precisely the right data, at the right level of quality, from the right sources, at the right time, to enable valid and timely decisions at minimum cost to the underlying network. This paper presents the decision model used and the protocol design philosophy, reviews the key recent results and describes a novel system, called Athena, that is the first to embody the aforementioned data delivery paradigm. Evaluation results are presented that compare the performance of decision-centric anticipatory information delivery to several baselines, demonstrating its various advantages in terms of decision timeliness, validity and network resources used. The paper concludes with a discussion of remaining future challenges in this emerging area.

Keywords: decision-centric Quality of Service (QoS); information delivery for the Internet of Things (IoT); real-time scheduling

---

1. Introduction

The concepts of decision-driven resource management, decision-centric information monitoring and decision support systems have been studied extensively in prior literature. As discussed later in the Related Work section, the system presented in this paper complements work on existing decision-centric monitoring and support tools and augments the general category of decision-driven resource management frameworks. It is distinguished by adapting decision-driven resource management to the realm of runtime network data management in the context of distributed sensing and Internet of Things (IoT) applications. This is as opposed to topics such as managing software architecture, human resources, financial assets or physical organizations. The application of decision-driven resource management to runtime networking leads to novel challenges and research questions in the design of network communication protocols and communication scheduling policies, presented in this paper.

This paper describes a novel model and system for anticipatory information delivery in mission-centric applications, where communication protocols and scheduling policies for data delivery are optimized specifically for meeting decision needs. We envision an environment such as an IoT-enabled smart city [1,2], where a cyber-physical system [3,4] is deployed to assist a team in performing a first-response mission in the aftermath of a natural disaster. Mission execution is affected by answers to a set of questions that first-responders might ask; for example: What are the affected areas? What is the extent of damage in each area? Who are the affected individuals? Where are the survivors? Are there injuries? Who is in most need of help? These and similar questions drive the decisions of the mission commander in allocating human and physical resources to different rescue efforts. A computing system familiar with similar missions can anticipate the types of decisions the commander must make, as well as the nature of mission-relevant information that affects these decisions. The system can then collect such information ahead of time and update it as the situation changes. In recent work, this information delivery paradigm was called decision-driven execution [5]. It re-thinks scheduling and communication to consider application models that are more pertinent to mission-driven quality-of-service-aware cyber-physical systems. We review this recent work, describe its embodiment in a system called Athena and present significantly expanded evaluation results that describe its advantages over a set of baselines.

Decision-centric anticipatory information delivery is motivated by visions of an instrumentation-rich world where a large number of sensors and other information sources may supply data relevant to an application [6,7]. There might not exist an a priori structure (such as fixed control loops) that statically dictates where each sensor should send its data. Rather, information sources will generally be connected to shared media, allowing flexible reconfiguration to support different (both one-off and repetitive) tasks. A challenge is to configure data collection from appropriate sources in order to meet specific application information needs while minimizing cost. For example, consider the myriad of sensing, storage and computing devices that comprise a megacity's cyber-infrastructure. Many sensors will be embedded in the physical environment. Different application tasks may need subsets of these devices. A sensor is activated when the user needs to perform a (decision) task that needs data from that sensor. In other words, resource consumption is decision-driven.

At a higher level, one can view decision-centric anticipatory information delivery as a way for data fusion systems to exert initiative in anticipating and offering mission-relevant information to mission commanders. For the system to exert initiative, it must know the nature of the mission (i.e., mission-relevant decisions that need to be made) and the unknowns that need to be determined in making mission-relevant decisions. The paradigm offers an exciting foundation for rethinking resource management as a process of arbitrating the acquisition and movement of data among processing components to meet decision needs. Such arbitration is guided by a novel set of questions: What data would be more relevant for making a decision? Which sensors are most appropriate for collecting such data? When should the data be collected to meet the freshness needs of the decision? What data should be cached in the distributed system and which nodes should store it to support aggregate decision needs most efficiently? At least three conditions must be met by the resource management algorithms; (i) the data collected for making a decision must be of sufficient quality to support the decision; (ii) it should be acquired sufficiently recently such that it is not stale by the time the decision is made; and (iii) the decision made based on collected data must meet the relevant decision deadline.

A distributed system that supports the decision-driven paradigm must optimize communication, storage and scheduling to meet these constraints. In this paper, we describe the implementation of one such system, called Athena, and evaluate its performance.

Decision-centric anticipatory sensor information delivery is an interesting cyber-physical problem. The data of interest typically come from sensors and as such captures aspect of the physical state of the world. Since the world state is dynamic, data objects have expiration time constraints, after which they become stale. The scheduling of data acquisition must obey these constraints. At the same time, decisions have deadlines, after which the window of opportunity to act will have passed. The system should therefore be cognizant of both the constraints arising from data freshness needs, as well as those arising from decision deadlines. Both sets of constraints are a function of the models of the physical world. Therefore, the combination of these sets of constraints leads to interesting new scheduling problems, where the intellectual innovations arise from simultaneously addressing requirements from the cyber realm (e.g., resource capacity constraints) and requirements from the physical realm (e.g., data freshness).

An abbreviated version of this work was presented at ICDCS 2017 [5]. This paper reviews the earlier work, places the effort in the broader category of decision-driven resource management frameworks, offers additional detail on the algorithms and implementation and significantly expands the evaluation results. Section 2 presents related work, placing decision-centric anticipatory information delivery in the broader context of decision monitoring and support tools and decision-driven resource management. Section 3 describes the new decision-driven system architecture for managing runtime data communication to optimize sensor information delivery. Section 4 overviews the resulting communication and scheduling challenges. Section 5 elaborates challenges in real-time scheduling that arise in the decision-driven context. Section 6 details networking challenges. The Athena implementation and evaluation are described in Sections 7 and 8, respectively. A brief discussion of other remaining research questions is presented in Section 9. The paper concludes with Section 10.

2. Related Work

Our work is broadly related to the concepts of decision-driven resource management, decision-centric information monitoring and decision support systems that have been studied extensively in prior literature in multiple domains. A significant amount work on decision-centric management revolves around managing software architecture decisions [8–10], product design [11], business processes [12], business intelligence [13] and physical organizations [14] (including human resources and financial assets). In contrast, we are the first to focus on decision-driven management of runtime communication and scheduling in networks that deliver sensory information. Below, we elaborate on this distinction, comparing our work more clearly to the general directions in current literature, including decision-centric management, decision-support tools and decision-driven information monitoring tools.

Much of today's decision-centric management frameworks are studied in disciplines such as business management [12,13] and organizational design [14]. They refer to human-centric processes that empower decision-makers, as opposed to runtime algorithms such as communication protocols and scheduling policies in computing machines. A category that comes closer to computer science in this context is the category of managing software architecture design [8–10]. Architectural decisions affect software performance outcomes. Hence, much work is focused on understanding the dependencies between decisions and outcomes, offering tools to enable decisions that lead to improved outcomes, such as improved performance [10], improved security threat mitigation [15,16] and improved defense [17]. For example, decision-driven architecture reviews [10] offer opportunities to assess the suitability of architectural decisions to design purposes early on the development cycle. Tools for decision-driven business performance management [18] document the relation between performance metrics on the one hand and decisions/sub-decisions on the other, allowing one to explore the decision space by modifying decisions and tracking their impact on performance metrics. These approaches are intended for use to inform design early on in the software development cycle (i.e., before deployment). In contrast, we are interested in communication protocols and scheduling policies that offer runtime support.

A somewhat closer tool category for our work is the category of decision support systems [19]. These systems use models of the world to allow decision-makers to play "what-if" scenarios and explore the consequences of their decisions on the modeled domain of interest. Hence, for example, given an adequate model of the political, military, economic, social and topological terrain features in some foreign state, a decision-maker might estimate the potential consequences of a certain military operation. Various planning tools have also been proposed that allow decision-makers to compose detailed plans of action and contingencies based on domain models [20]. Work reported in this paper does not constitute a decision-support tool in the above sense. The software described in this work does not have models of the world and is unable to compose plans. Rather, it helps users to retrieve information objects in a manner that facilitates meeting decision needs.

The concept of decision-centric information monitoring comes closest to our work. A decision-centric information monitoring system [21] answers the following key question: which variables must one monitor in order to make an informed decision? The answer often comes from modeling the decision-making process [13,22]. An interesting trade-off is involved. Monitoring too many variables may overwhelm the decision-maker, whereas monitoring too few may impair the quality of decisions. This work is synergistic with ours in that we assume that the problem of determining the relevant variables (or unknowns) has already been solved. Cast in the context of a cyber-physical systems in which unknowns that determine the viability of each potential course of action have been identified, our work addresses the complementary problem of contacting the sources (e.g., sensors) that would deliver information on these unknowns over a network while minimizing delivery cost. This network communication and scheduling policy challenge is orthogonal to the manner in which relevance of specific unknowns was determined in the first place.

Recent work has made initial progress at solving the above cyber-physical network resource management problem [5,23–25]. The work bears resemblance to prior database research that considered explicit data access transactions and required a degree of data freshness [26–29]. It is new in tying such retrieval policies explicitly to the decision needs, leading to new research challenges and solutions as described below.

3. A Decision-Driven System Architecture

A data fusion system that supports decision-centric anticipatory information delivery may simultaneously retrieve data for multiple missions. An example mission might be to track a specified target (such as a robbery getaway vehicle or an escaped fugitive) in a city using deployed sensors and security cameras or to rescue a group of individuals trapped in the aftermath of a natural disaster. Execution of the mission starts with collecting information relevant to the mission. We assume that such collection occurs over a tactical network of limited resources, which is often the case in such missions as disaster response (where much infrastructure has been destroyed, leading to resource shortages). The primary purpose of the data collection system is to conserve the resources expended on data collection while at the same time collecting enough data to support the mission commander or decision-maker. In the context of a mission, the commander or decision-maker needs information to make decisions. In a rescue mission, an example decision might be: Which is evacuation path to follow in order to rescue the disaster survivors? In a tracking mission, a decision might be: Which sensors are to be turned on in order to track the target? An interesting trade-off is involved. Namely, collecting more information may overwhelm the underlying resource-constrained network, leading to delays that may negatively impact the ability of the user to obtain information on time and hence their ability to make timely decisions. In contrast, collecting inadequate information might impede one's ability to make a correct decision. For example, if one does not collect enough information on the health conditions of various roads in the aftermath of a large natural disaster, one might attempt to evacuate the survivors over a road that ends up being blocked or otherwise unsuitable. The key is therefore to collect just the right amount of information to make a decision. With that in mind, several levels of indirection allow the system to optimize information delivery for the respective missions, as follows:

- The decision working set: For each mission, the system maintains the set of most common decisions to be made, called the decision working set. For example, in a rescue mission, decisions may need to be made on the best evacuation route for each survivor. The set of all such mission-relevant decisions constitutes the working set.

- The decision model: In deciding on a course of action, the commander or decision-maker must consider several relevant variables that impact the decision outcome. For a trivial every-day example, in deciding what to wear in the morning, one might consider weather conditions (such as temperature and precipitation). These variables constitute the unknowns that need to be determined for a decision to be made. Hence, for each decision in the working set mentioned above, the system must know the relevant unknowns, as well as how these unknowns impact decision outcome. We call it the decision model. The choice of decision model itself gives rise to interesting research questions that warrant further attention. In the simplest model, decisions are viewed as choices of a course of action among multiple alternatives [5,25]. More on the decision model will be mentioned later in the paper. The viability of each individual alternative depends on the satisfaction of several predicates. Making a choice can therefore be thought of as an evaluation of a logical expression of multiple predicates; for example, "if it is (i) sunny and (ii) warm, I will wear a T-shirt; else, I will wear a sweater".

- The unknowns: Consider a predicate such as "if it is (i) sunny and (ii) warm". Evaluating such a predicate requires determination of the value of one or more unknowns. In the example quoted above, the unknowns are parameters of local weather that determine whether or not it is sunny and whether or not it is warm. The decision model specifies the unknowns whose value needs to be determined in order to decide on the viability of each course of action.

- The evidence data objects (or simply, data objects): Determination of the unknowns entails the acquisition of corresponding evidence. Data objects such as images, videos or sound clips, generated by appropriate sensors, can supply the needed evidence. For example, a picture taken several minutes ago at the location of interest, showing that it is sunny, would constitute evidence that "sunny" can be evaluated as "true".

- The sources: Often, a piece of evidence (e.g., a picture that shows whether it is sunny or rainy) can be supplied by any of several alternative sources, such as multiple cameras overlooking the scene. The system must choose a source such that timely and relevant information is provided at low cost. This gives rise to appropriate source selection and data delivery scheduling protocols.

Figure 1 is a conceptual view of the architecture, where missions are broken into decision working sets, decisions are associated with logical expressions in appropriate unknowns and the unknowns are linked to sources that can supply the relevant data objects (or evidence).

A decision-driven resource management system allows applications to make queries we call decision queries, or decision tasks, that request information needed for a decision. The system comprises nodes that contribute, request or help forward data needed for these decisions. It manages the acquisition of evidence needed to evaluate the viability of different courses of action involved in decision-making. By accounting for models of decisions and sources, the system carries out the required information collection and transmission in a more efficient and timely manner to support decision-making.

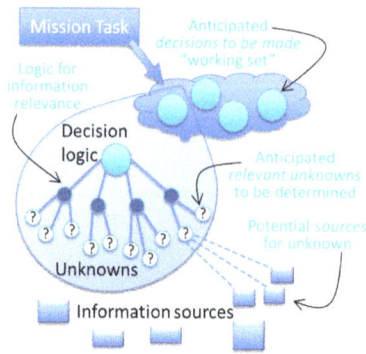

Figure 1. A conceptual architecture for anticipatory information delivery.

Athena is a recently developed system that implements the above architecture. An illustration of Athena system components is shown in Figure 2. For completeness, we list all components of the overall system below. Later in the paper, we shall focus only on data communication and scheduling policies and the components that impact them. Figure 2 presents the overall architecture. The following components are depicted:

- User I/O: This is the user interface component that allows entry of decision queries and missions. An instance may run on every Athena-enabled node in the distributed system. The interface passes user queries to the rest of the system. It may also originate anticipated queries based on the mission's decision working set.
- Application semantic translator: This component determines the unknowns needed for a particular decision. These unknowns are mission specific. A mission-specific library implements the application semantic translator. In the current system, libraries have been implemented for tracking missions and rescue missions, as a proof of concept.
- Semantic store: This component maps individual unknowns to sources who may have evidence objects that determine the value of that unknown. The semantic store may be replicated. Sources would send metadata about objects they have to the semantic stores that allow the latter to match unknowns and sources.
- Source selector: It is often the case that multiple sources have redundant information and hence can help resolve the same unknown. It would be wasteful to collect data from all these redundant sources. Instead, the source selector determines which subset of data sources to contact.
- Logical query resolution engine: This engine determines the order in which evidence objects are to be acquired from their sources to ensure constraints such as data freshness and decision deadlines.
- Information collection and dissemination engine: This engine offers the mechanisms for data collection, including mechanisms for prefetching potentially relevant data at a lower priority. This component will be described later in the implementation section in more detail.

A user's query (actual or anticipated) is accepted and translated into the corresponding logical expression (predicates over relevant unknowns) by the Application Semantic Translator , which then uses a nearby Semantic Store to identify the set of nodes that my have evidence objects to determine the unknown. Taking this information, the Logical Query Resolution Engine then uses the Source Selector to choose some subset of these sources to contact. This subset aims to minimize redundancy of the contacted node set, as well as delivery cost. Data requests are then scheduled in an order that takes into account decision deadlines and data freshness constraints. These requests are handed over to the Information Collection & Dissemination Engine, which accesses the underlying sensing and communication stacks to handle all incoming and outgoing requests, data objects and resolved predicate label values, with proper book-keeping and information updates to its various internal components, as discussed in detail later.

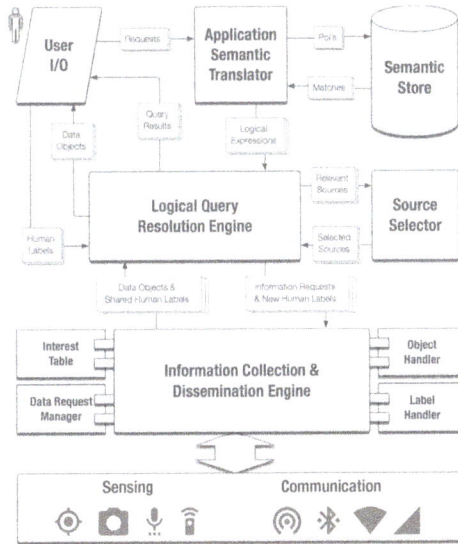

Figure 2. Athena's architectural design, as well as how information propagates through different system components.

3.1. Exploiting Decision Structure: An Illustrative Example

A key innovation of the decision-driven system lies in a novel query interface that allows applications to express decision needs in a manner that helps the resource management components properly prioritize data acquisition. Specifically, a query may specify a logical expression that describes the decision structure. This expression specifies the predicates that need to be evaluated for the corresponding choice (of a course of action) to be made. In this model, there are no limits on the types of queries that can be expressed as long as they can be represented by Boolean expressions over predicates that the underlying sensors can supply evidence to evaluate.

There are many possible ways that such expressions could be obtained. In many applications, especially those involving liability or those where human teams must operate efficiently under adverse or dangerous conditions, a well-prescribed operation workflow is usually followed. The workflow specifies how individuals should act, under which conditions a given course of action is acceptable, and what checks must be done before embarking on an action. Training manuals, rules of engagement, doctrine, standard operating procedures, and similar documents describe these workflows, essentially documenting acceptable decision structures. Decision logic could also be learned by mining datasets that describe conditions observed and decisions taken on them by an authority. Such an approach, for example, may be used to reverse-engineer strategy used by an expert or by an adversary. Finally, in some cases, decision logic could be algorithmically derived. For example, in a vehicular navigation application, the driver will generally seek a route that satisfies some machine-checkable property, such as a condition on expected commute time, quality of route, or length of commute. Hence, the logic for the decision on route from alternatives on a given map is known. An interesting research question is: given the logical decision structure (i.e., the graph of logical predicates to be evaluated to arrive at a course of action), how best to deliver the requisite information?

Let us look at a toy example to help make the picture more concrete. Consider a scenario from emergency response [30]. Suppose after an earthquake that hits our smart city, there is a shortage of air support, and an emergency medical team needs to transport a severely injured person from an origin site to a nearby medical center for surgery. There are two possible routes to take: One composed of segments *A–B–C*, and the other of segments *D–E–F*. We need to make sure that the chosen route is in good enough condition for our vehicle to pass, so we want to retrieve pictures from deployed roadside cameras in order to verify the road conditions and aid our decision-making on which route to take. Our route-finding query can be naturally represented by the logical disjunctive norm form $(viable(A) \wedge viable(B) \wedge viable(C)) \vee (viable(D) \wedge viable(E) \wedge viable(F))$, where $viable(X)$ represents the predicate "route segment *X* is viable". This expression signifies that at least all segments of one route need to be viable for the transport to occur. In this example, if road segments *A*, *B* and *C* all turn out to be in good condition, then the first route is viable, and there is no need to continue retrieving pictures for road segments *D*, *E*, and *F*. Similarly, if a picture of segment *A* shows that it is badly damaged, we can skip examining segments *B* and *C*, as this route is not going to work anyway. Instead, we can move on to explore segments *D*, *E*, and *F*.

As is evident from this toy example, exploiting decision structure (represented by the Boolean expression) enables us to take inspiration from heuristics for short-circuiting the evaluation of logical expressions to schedule the acquisition of evidence. Specifically, we can acquire evidence in an order that statistically lowers expected system resource consumption needed to find a viable course of action. By incorporating additional meta-data (e.g., retrieval cost of each picture, data validity intervals, and the probability of each road segment being in good or bad condition), we can compute retrieval schedules that better optimize delivery resources expended to reach decisions. This optimization, indeed, is the main research challenge in the decision-driven execution paradigm.

This optimization must consider both physical and cyber models. On one hand, models of the underlying physical phenomena are needed to correctly compute inputs such as data validity intervals (how long can one consider measurements of given physical variables fresh), and environmental conditions (e.g., probabilities that some measurements not yet acquired will fall into a range that invalidates versus supports a predicate). One the other hand, models of computing and communication resources are needed to understand how much bandwidth and compute power are available for data collection from the physical world.

The latter models can be obtained from network and other resource monitoring. The former are more difficult to obtain. They can be learned over time or derived from the physical nature of the phenomena in question. For example, temperature does not change very quickly. Hence, the validity interval of a temperature measurement could be of the order of large fractions of an hour. On the other hand, state during an active emergency, such as a burning building, can change on the order of minutes. Hence, its validity interval is much shorter. It is also possible for external events to invalidate freshness of variables. For example, the existence of a resource, such a bridge across a river, can be assumed to hold with a very large validity interval. However, a large earthquake or a military air-raid may invalidate such past observations, making them effectively stale and in need of being re-acquired from sensors. The same applies to learned probabilities of conditions. The probability of traffic congestion on some freeway at 11 p.m. on a Monday night might be known. However, a condition, such as a nearby large concert that ends around the same time, can invalidate it. In general, a combination of past contextual knowledge, current observations, and invalidation will be needed to operationalize the physical models.

Lowering the data acquisition costs of decisions involves carrying out an optimal collection strategy given the resources available and the underlying physical models, such that a measure of decision correctness is maximized, while cost is minimized. If some contextual information needed for the models is not known, the optimization may proceed without it, but the quality of solutions will be lower, generally entailing a less than optimal resource cost. The sensitivity of decision cost to the quality of models supplied is itself an interesting research problem.

3.2. System Abstractions and Components

The decision-driven execution system represents the physical world by a set of labels (names of Boolean variables). These labels can be used in expressions of decision logic structures. The system maintains tuples of (*label, type, value*), where *label* is just an identifier (i.e., variable name), the *type* specifies the semantic type of the label (for example, "road condition"), and *value* could be true, false, or unknown. The system can be easily extended to more general types (other than Boolean). More general discrete variables can be implicitly represented by sets of labels, one label for each allowed value of the variable, with the restriction that only one of these can be true at a time. Continuous variables can be supported as long as actions are predicated on some thresholds defined on these variables. For example, the decision to turn the lights on in a smart room can be predicated on the value of an optical sensor measurement dropping below a threshold. This is a Boolean condition whose evaluation result can be stored in a variable labeled, say, *Dim*. The pool of labels itself can be dynamic. New applications can add new labels (and new categories of labels) to the pool and specify sensing modalities needed to determine label values. For instance, in the routing example above, the predicate *viable*(*X*) can be represented by the label *viableX*, denoting a Boolean variable of value true (if the route segment is viable) or false (if it is not). The route selection decision is associated with labels *viableA*, *viableB*, ..., *viableF*.

To determine the value of a label (e.g., whether conditions of a road segment make it a viable candidate), evidence must be collected. An example of such evidence might be a picture of the corresponding road segment. We call such evidence items evidence objects or simply data objects, where it is clear from context that the data in question offers evidence needed to evaluate a logical predicate in the decision structure.

Evidence objects are data objects needed for deciding the value of labels. Entities that examine evidence in order to determine the value of a label are called, in our architecture, annotators. For example, an annotator could be a human analyst receiving a picture of route segment *A*, and setting the corresponding label, *viableA*, to true or false, accordingly. Alternatively, an annotator could be a machine vision algorithm performing the same function. In general, annotators should advertise the type of evidence objects they accept as input, and the types of labels they can accordingly compute. Clearly, the same object can be used to evaluate several different labels. For example, a picture of an intersection can be used to evaluate physical road conditions. However, it can also be used to detect specific objects such as individual vehicles, license plates, or pedestrians, or used to estimate values such as length of traffic backup, traffic speed, or congestion level.

Another key component of the decision-driven resource management paradigm is the data sources. Sources that originate data, such as sensors, must advertise the type of data they generate and the label names that their data objects help resolve. For example, a source might offer pictorial evidence of road conditions. Such a source would advertise both its data type (say, JPEG pictures) and the specific geographic locale covered. In the route discovery example, this source would need to be paired with an annotator that can accept pictures as input and determine viability of road segments within that geographic locale.

Finally, an important component is network storage or caches. The decision on mapping data and computation to network nodes in a distributed execution environment is a classical problem in distributed computing systems. This problem must be solved in the context of a decision driven execution as well. Content (both data objects and annotation labels) should be cached at nodes closer to consumers who might need these objects and labels for their decision-making. Similarly, annotators will need to execute on nodes that are close to consumers needing the annotations. The placement of data and computational modules in the network to minimize decision cost remains an open problem.

The aforementioned architecture effectively changes the query paradigm from specifying what objects to retrieve to specifying why they are needed; that is to say, how they fit in the logic used to make a decision. This shift is thanks to sharing the structure describing the query's decision logic. Evidence objects are needed to resolve predicates named by labels in that decision logic. The architecture allows the network to be much smarter when answering a query. Being aware of the logical decision structure, the resource management system can allocate resources to seek evidence that helps evaluate the decision expression at the lowest cost. As alluded to in the introduction, we can take inspiration from literature on optimizing the evaluation of logical expressions to determine which labels should be evaluated first and which sources should be contacted for the corresponding evidence. In turn, this determination informs resource allocation, such as policies for scheduling/queuing of object retrieval requests, policies for caching of results, and choices governing invocation of annotators.

3.3. A Walk through the Execution of a Decision Query

Putting it all together, when a user makes a decision query, at a high level, query resolution works as follows. The system first determines the set of predicates (i.e., labels) that is associated with the query from the underlying Boolean expression that describes the decision logic. This is the set of labels whose values need to be resolved. The query source then needs to determine the set of sources with relevant evidence objects. If multiple sources offer redundant evidence, some arbitration is needed to determine who to contact. A scheduling algorithm must decide on the order in which evidence objects must be retrieved to evaluate the different labels.

The system must manage caching. Say, the query source decides to resolve the value of the label, *viableX*. If the label has already been evaluated in the recent past (because of a prior query), its evaluation may be cached in the network, in which case the resolved value can be found and returned. This is the cheapest scenario. Otherwise, if the evidence object needed to evaluate the predicate has been recently requested (but the corresponding label not evaluated), the requested object may be cached. Such might be the case, for example, when the object was requested to evaluate a different predicate. The cached object needs to be sent to the right annotator to determine the label value relevant to the current query. Otherwise, if the objects is not cached or is stale, the query should be propagated to a source that has fresh relevant objects. The relevant object is then shipped to an annotator that decides label values. Both the object and the computed new labels are cached in the network with a freshness interval that specifies their validity for future use. Next, we outline the research challenges that must be addressed in realizing this architecture.

4. Decision-Driven Resource Management: Optimizing Retrieval Cost

Initial work on decision-driven resource management was recently published in the context of centralized systems [25,31]. It needs to be extended to a more general decision model and to distributed resource management. Consider a workload model, where tasks consume resources to make decisions, each represented by a logic expression in disjunctive normal form (OR of ANDs). Let $\{a_i\}$ denote the set of alternative courses of action for the *i*-th decision, and $\{b_{i_j}\}$ denote the *j*-th Boolean condition needed to determine the value of a_i. Therefore, a query *q* takes the general form:

$$q = \underbrace{(b_{0_0} \wedge b_{0_1} \wedge \ldots)}_{a_0} \vee \underbrace{(b_{1_0} \wedge b_{1_1} \wedge \ldots)}_{a_1} \vee \ldots .$$

The first challenge lies in designing algorithms that optimize the cost of retrieving evidence objects needed to resolve the decision query. In the simplest model, the query is resolved when a single viable course of action is found. Other more nuanced models may be possible. For example, a query could be resolved when a viable course of action is found for which additional conditions apply that may be represented by another logical expression structure ANDed with the original graph.

4.1. Minimizing Retrieval Cost by Short-Circuiting

Associated with each condition b_{i_j} may be several pieces of metadata. Examples include (i) retrieval cost C_{i_j} (e.g., data bandwidth consumed), (ii) estimated retrieval latency l_{i_j}, (iii) success probability p_{i_j} (i.e., probability of evaluating to true), and (iv) data validity interval d_{i_j} (i.e., how long the data object remains fresh). The question becomes: how to orchestrate the retrieval such that the query is resolved at minimum cost?

Sequential retrieval of evidence objects gives the most opportunity to take advantage of the decision logic structure to short-circuit and prune unnecessary retrievals in view of previously retrieved objects. Simply put, when handling an AND,

$$a_i = b_{i_0} \wedge b_{i_1} \wedge b_{i_2} \wedge \dots,$$

we want to start with the most efficient b_{i_j} and proceed downwards. Here, "most efficient" means highest short-circuit probability per unit cost:

$$\frac{1 - p_{i_j}}{c_{i_j}}.$$

Imagine a particular course of action whose viability depends on just two conditions, h and k, that require retrieving and examining a 4-MB and a 5-MB audio clip, respectively. It has been estimated (e.g., from historic data or domain expert knowledge) that condition h has a 60% probability of being true, whereas k has a 20% probability. In this case, we would want to evaluate k first, as it has a higher short-circuiting probability per unit bandwidth consumption. Intuitively, this is because it is more likely to be false, thereby producing a result that obviates retrieval and evaluation of the remaining ANDed primitives. More precisely:

$$\underbrace{\frac{1 - 0.2}{5}}_{0.16} > \underbrace{\frac{1 - 0.6}{4}}_{0.1}.$$

Hence, this evaluation order leads to a lower expected total bandwidth consumption compared to the other way around (i.e., evaluating h before k):

$$\underbrace{5 + 0.2 \times 4}_{5.8} < \underbrace{4 + 0.6 \times 5}_{7}.$$

Similarly, for the handling an OR in the logic structure:

$$q = a_0 \vee a_1 \vee a_2 \vee \dots,$$

we start processing the a_i with the highest short-circuiting probability per unit cost; in this case, one that has the highest probability of evaluating to true.

Conditions in the physical world can change over time. Therefore, it is important that, at the time a decision is made, all pieces of information involved must still be fresh. Otherwise, decisions will be made based on (partially) stale information. A greedy algorithm has been proposed [25], where all data object requests are first ordered according to their validity intervals (longest first) to meet data expiration constraints, then rearrangements are incrementally added, according to objects' short-circuiting probabilities per unit cost, to reduce the total expected retrieval cost.

The approach is heuristic and does not have a known approximation ratio. Near optimal algorithms should be investigated. Unlike early work that considers object retrieval over a single channel, it is interesting to extend the formulation to consider more general network topologies. Importantly, this retrieval order is influenced by models of the physical world that determine how fast physical state changes, and thus how often it needs to be sampled. Such models will be incorporated into the optimization to refine expressions of short-circuit probability. Specifically, whether or not a retrieved object short-circuits an expression depends not only on the value of the corresponding predicate evaluation, but also on when the evaluation was carried out. Stale evaluation results are not useful. Hence, the optimization must be cognizant of timing constraints derived from physical models of the underlying measured phenomena. The complete algorithm pseudo-code is shown in Algorithm 1.

Algorithm 1 Retrieval schedule for dynamic query resolution.

Input: A query's deadline requirement d_q, its candidate courses of action $\{a_i\}$, and acceptable parallel retrieval level r. For each constituting condition t_{i_j} for a particular a_i, its corresponding

- evaluation (retrieval) costs c_{i_j},
- retrieval latencies l_{i_j},
- success probabilities p_{i_j}, and
- freshness (validity) interval v_{i_j}.

Finally, the subset of conditions that have been cached O_q, in descending order of $\frac{1-p_{i_j}}{c_{i_j}}$.

Output: Query resolution result

1:   $failure\_counter \leftarrow 0$
2:   $L_a \leftarrow \{a_i\}$ sorted in descending order of $\frac{p_i}{c_i}$
3:   **for** a_i in L_a **do**
4:     $Q_c \leftarrow \varnothing, Q_d \leftarrow$ longest valid. interval first order, $S_p \leftarrow \varnothing$
5:     $L \leftarrow \{t_{i_j}\}$ sorted in descending order of $\frac{1-p_{i_j}}{c_{i_j}}$
6:     **while** $Q_d \neq \varnothing$ **do**
7:       **for** t_l in L **do**
8:         **if** moving t_l from Q_d to the end of Q_c does not increase freshness violation degree **then**
9:           $Q_d \leftarrow Q_d \setminus \langle t_l \rangle, Q_c \leftarrow Q_c + \langle t_l \rangle$
10:           **break**
11:         **end if**
12:       **end for**
13:     **end while**
14:     **while** $|Q_c| > 0$ **do**
15:       $t_e \leftarrow$ end element of Q_c
16:       $Q_c \leftarrow Q_c \setminus \langle t_e \rangle, S_p \leftarrow S_p \cup \{t_e\}$
17:       **if** $Q_c + S_p$ satisfies validity intervals **then**
18:         **for** t^o in O_q **do**
19:           $d_{t^o} \leftarrow t^o$'s absolute validity deadline
20:           **if** $Q_c + S_p \setminus \langle t^o \rangle$ satisfies $\min(d_q, d_{t^o})$ **then**
21:             $Q_c \leftarrow Q_c \langle t^o \rangle, S_p \leftarrow S_p \setminus \langle t^o \rangle$
22:           **else if** a shortest tail, T_{Q_c} of Q_c's can be moved to S_p to satisfy validity intervals **AND** $|S_p| \leq r$
23:             $Q_c \leftarrow Q_c \setminus T_{Q_c}, S_p \leftarrow S_p \cup T_{Q_c}$
24:             update d_q
25:           **end if**
26:         **end for**
27:         Process a_i with retrieval schedule $Q_c + S_p$
28:         **if** a_i succeeds **then**
29:           **return** a_i as an successful result
30:         **else**
31:           increment $failure\_counter$ by 1
32:           **break**
33:         **end if**
34:       **end if**
35:     **end while**
36: **end for**
37: **if** $failure\_counter == |L_a|$ **then**
38:   **return** request resolves to failure
39: **else**
40:   **signal** validity interval cannot be satisfied
41: **end if**

4.2. Minimizing Retrieval Cost by Optimizing Coverage

Another interesting question in minimizing the cost of object retrieval lies in selecting the sources from which objects should be retrieved, as well as the annotators needed to compute predicate values from the supplied evidence. Three interesting challenges arise in the context of this optimization.

First, in general, multiple sources may offer evidence objects that help evaluate the same or overlapping subsets of predicates needed for resolving a decision query. Some evidence objects may lead to evaluating multiple predicates at once. In our running example of route finding, a single picture from an appropriate camera can help evaluate conditions on multiple nearby road segments at once, if all such segments are in the camera's field of view. Hence, to determine the most appropriate sources to retrieve evidence from, one must solve a source selection problem. This problem can be cast as one of coverage. It is desired to cover all evidence needed for making the decision using the least-cost subset of sources. Variations of this problem will be investigated in the proposed work.

Second, an interesting novel factor in our resource management model is the existence of annotators. Not only do we need to collect evidence objects, but also we want to use them to determine specific predicate values. As mentioned earlier, an annotator could be a human, in which case one must consider the cost of delivering the collected evidence to that human for annotation. Alternatively, the annotator could be a machine. When the annotator is the query source, all evidence must simply be shipped to that source for both annotations and decision-making. In this case, we assume success at resolving the query as long as all evidence objects can be shipped by the decision deadline and remain fresh at that deadline. When the annotator is a piece of software, we other challenges arise. For example, where in the distributed system should that software be located to minimize decision cost? Besides considerations of network cost, how to account for processing factors such as load balancing on the annotators?

Finally, there is the issue of confidentiality and trust. A user might not trust the accuracy of specific annotators or might not wish to send specific evidence objects to them for confidentiality reasons. Such additional constraints will be incorporated into the optimization algorithm. To address trust, the label values computed by different annotators will be signed by the annotator. Such signatures can be used to determine if a particular cached label meets the trust requirements of the source. Similarly, labels can note which objects the annotator used to make their annotation decision. That way, trust becomes pairwise between the annotator and the source. If an annotator requires multiple pieces of data to solve a predicate, then all are stored in the label. In JSON, one can think of the following label format:

```
{
  "label":"viableX",
  "type":"road condition",
  "value":true,
  "annotator":"/BBN/boston/bldg9/photo_analysis_v2.39",
  "sources": ["/city/marketplace/south/noon/camera1",
     "/city/marketplace/north/dawn/camera5"]
}
```

5. Real-Time Decision-Driven Scheduling

The architecture described in the previous section inspires opportunities to develop a new type of real-time scheduling theory, we call decision-driven scheduling. The objective of a decision-driven scheduling algorithm is to schedule the retrieval of data (evidence) objects needed for current decision queries.

As mentioned earlier, two types of timing constraints must be enforced by the retrieval schedule. First, decision deadlines must be obeyed (i.e., deadline constraints). Second, data furnished for a decision must be fresh (i.e., data validity constraints). Each retrieved data object has a validity interval within which it remains fresh. A decision is valid only if it is made based on objects that remain within their validity intervals.

5.1. Initial Results

Simple versions of the above problem have been solved in recent work [23,24]. For example, consider the basic case of a single task (i.e, decision query) deciding the viability of a single course of action, where the underlying data objects are retrieved over a single resource bottleneck. For simplicity, let the source also be the data annotator. Hence, the system must simply deliver all evidence objects to the source by the decision deadline.

In this scenario [23], let there be N objects, named $O_1, \ldots, O_N$, that have information relevant to the decision. These objects are retrieved from sources identified by the source selection algorithm. Let us denote these sensors by $S_1, \ldots, S_N$, respectively. Let us further consider that these sources monitor their environment on demand. Hence, when source, S_i, is contacted, it observes its environment and constructs an appropriate observation object, O_i (e.g., takes a picture), then sends it in response to the received request. Let us assume that O_i has a validity period I_i and size C_i. Let source S_i be contacted at time t_i.

The system collects the relevant data from all selected sources then a decision is made. Let time F denote the time when all data needed for the decision has been fetched, such that a decision can be made. We require that $F \leq D$, where D is a deadline. If all N objects are retrieved, the decision cost is given by:

$$Cost_{opt} = \sum_{1 \leq i \leq N} C_i \tag{1}$$

An interesting question is: does there exist a feasible object retrieval schedule? In other words, is it possible to fetch all N objects such that the decision deadline is met (to make sure the decision is timely) and such that each object remains fresh by the time fetching is complete? The latter ensures that the decision is valid (because it was made based on unexpired data). This occurs when no sensor is sampled twice. Let a feasible retrieval schedule be one that satisfies the decision deadline. An optimal retrieval scheduling policy is one that finds a feasible retrieval order whenever one exists, thereby simultaneously satisfying both freshness and deadline constraints:

Data freshness: $t_i + I_i \geq F$ $(\forall i, 1 \leq i \leq N)$,

Decision deadline: $t + D \geq F$,

where the decision query arrives at time t. The freshness constraint above ensures cost minimality. If it is violated, a second sample is taken from the sensor, which makes the cost non-optimal. These can also be represented together as:

$$\min \left(\min_{1 \leq i \leq N} (t_i + I_i), \ t + D \right) \geq F$$

Prior work [23] shows that the Least Volatile object First (LVF) object retrieval policy is optimal in the case of a single decision query where data are retrieved over a single bottleneck resource. The policy fetches the object with the longest validity interval first. The optimal policy for multiple independent decision queries is more involved, but has been derived for a simple decision model [23].

5.2. Remaining Challenges

The above work has several limitations. First, while it does consider multiple decision queries, they are assumed not to overlap in the sets of data objects they need. Second, the decision for each query involves evaluation of validity of only a single course of action. Hence, there is no disjunction in the decision model. Short-circuit opportunities are not considered. Finally, all objects needed for making the respective decisions are assumed to be retrieved over a single channel, essentially reducing the problem to one of single-resource scheduling. To establish a general theory of decision-driven scheduling these limitations need to be removed. This leads to several avenues of investigation:

- Non-independent queries: It is important to consider the case where some queries overlap in needed data objects. In this case, retrieving each object once is not optimal anymore. That is because, if an object is shared by multiple queries, there is a possibility that the same data object can be reused. Such reuse can reduce total cost. At present, the optimal solution to this problem is unknown. Algorithms with near optimal performance are needed. They should be further extended to account for more complex decision models (i.e., multiple courses of action) and short-circuit opportunities.
- Noisy sensor data: The challenge here is to adapt prior algorithms to the case where sensor data is not clean. Hence, it might not be enough to retrieve a single piece of evidence to evaluate some label. Rather, multiple pieces may be needed to corroborate the computed label to a specified degree of confidence. The need for such corroboration has implications on source selection and data retrieval schedules. Requirements for confidence in computed predicate values, in the presence of noisy data, lend themselves nicely to the formulation of new scheduling problems, where the right amount of evidence must be retrieved to guarantees a level of confidence in decision results. Annotators, in this scenario, may need to examine multiple pieces of evidence (e.g., multiple pictures) to determine the value of a particular label (e.g., whether a route segment is viable). Once the label value is determined, annotators can offer feedback on the quality of individual inputs used. For example, they may mark a given picture (and hence, its source) as not useful. Such feedback can accumulate to gradually build profiles for reliability of sources. In turn, these profiles may be considered in future source selection problems to avoid bad sources or seek sufficient corroboration such that a required level of confidence in results is attained. The problem gets more complicated by considering reliability of annotators. A bad annotator could offer false feedback that improperly influences the reliability profile of a source. Hence, individual query originators may develop different profiles for the same data sources, depending on which annotators they trust.
- Event-triggered decision-making: In many scenarios, the need for decision-making itself will be triggered by sensor values. For example, the firing of a motion sensor inside a warehouse after hours may trigger a decision task to determine the identity of the intruder. Other decisions may need to be done periodically. The scheduling problems described above can thus be augmented by analysis that takes into account decision triggers, offering a better model of expected future workload, such as periodicity, or specific contexts in which the decision query will arrive.

6. Network Challenges

In a distributed system where decision tasks can originate at different nodes and where evidence needed to make a decision may be distributed, it is important to address the underlying networking challenges. Specifically, how do we find sources who have evidence pertaining to the decision? Where to cache objects as they are retrieved from those sources? When objects are processed by annotators to generate values for one or more labels, where should these values be stored? Answers to these questions are needed in the context of three mechanisms, below.

6.1. Hierarchical Semantic Naming and Indexing

Since decision-driven resource management is centered around data retrieval, it seems natural that some form of information-centric networking can be implemented to facilitate routing queries and finding matching objects [32,33]. In information-centric networks, such as NDN [34], data, not machines, are the primary named entity on the network. The network adopts hierarchical data names, instead of hierarchical IP addresses. In this paradigm, consumers send low-level queries, called interest packets, specifying a data name or name prefix. Routing tables directly store information on how to route interests to nodes who previously advertised having data matching a name prefix. Hence, interests are routed directly to nodes that have matching data. The data then traverses the reverse path of the interest to return to the query originator.

Adaptations of the information-centric networking ideas can furnish the underlying framework for routing queries to sources in the decision-driven execution architecture. In an NDN-like implementation, evidence objects, labels, and annotators all have public names in an overall name space. Nodes possessing those objects advertise their names. Nearby routers who receive those advertisements update their tables such that interests in the given names are correctly forwarded to nodes that have matching objects. Since labels encode the semantics of the underlying variables, we call the resulting scheme hierarchical semantic indexing.

In designing hierarchical name spaces (where names are like UNIX paths), of specific interest is to develop naming schemes where more similar objects have names that share longer prefixes. This naming scheme will allow the network do clever object substitutions, when approximate matches are acceptable. For example, when a query arrives for an object /city/marketplace/south/noon/camera1/, if retrieving this object is impossible or costly, the network may automatically substitute it with, say, /city/marketplace/south/noon/camera2/. This is because the large shared name prefix signifies that the latter object is very similar to the former (e.g., a view of the same scene from a different angle). Hence, it is a valid substitution when approximate answers are allowed. This mechanism may lead to substantial resource savings and more graceful degradation with overload. In fact, it may offer a new foundation for network congestion control, where requirements on the degree of acceptable approximation are relaxed as a way to combat congestion and tightened again when congestion subsides.

6.2. Information-Maximizing Publish-Subscribe

Building on the aforementioned hierarchical semantic indexing, it becomes possible to develop network resource management protocols that maximize information flow from sensors to decision tasks. The importance of delivering a piece of information is not an absolute number, but rather depends on other information delivered. For example, sending a picture of a bridge that shows that it was damaged in a recent earthquake offers important information the first time. However, sending 10 pictures of that same bridge in the same condition does not offer 10-times more information. Indeed, the utility of delivered information is sub-additive. This observation has two important implications; namely:

- Data triage cannot be accurately accomplished by assigning static priorities to data packets, as the importance of one piece of information may depend on other information in transit.
- Data triage cannot be accurately accomplished at the data source, as the source may be unaware of other sources supplying similar information.

The above two points argue for implementing data triage in the network. An information-utility-maximizing network must perform data triage at network nodes to maximize the delivered (sub-additive) information utility in the face of overload. Our premise is that a network that explicitly supports hierarchical names for data objects (as opposed to hierarchical IP addresses for machines) can directly maximize and significantly improve delivered information utility. In a well-organized hierarchical naming scheme, objects with hierarchical names that share a longer prefix are generally closer together in some logical similarity space. Assuming that items closer together in that space share more information in common, distances between them, such as the length of the shared name prefix, can be leveraged to assess redundancy in sub-additive utility maximization. Since content names are known to the network, fast greedy sub-additive utility maximization algorithms can be implemented on links and caches. For example, the network can refrain from forwarding partially redundant objects across bottlenecks; it can cache more dissimilar content, and can return approximate matches when exact information is not available. The above intuition suggests that naming data instead of hosts lays a foundation for information utility maximization and for improving network overload performance.

6.3. Support for Different Task Criticality

Importantly, network resource management mechanisms must support tasks of different criticality. In a network that directly understands content names, it is easy to implement different content handling policies that depend on the content itself. Some parts of the name space can be considered more critical than others. Objects published (i.e., signed) by an authorized entity in that part of the name space can thus receive preferential treatment. There objects, for example, can be exempt from the aforementioned approximation mechanisms for congestion control. They can also receive priority for caching and forwarding. The integration of such preferential treatment mechanisms with the scheduling problem formulation described earlier is itself an interesting research problem.

7. Implementation

To perform a proof-of-concept validation, we implemented a distributed system, called Athena, that embodies the decision-driven execution paradigm. At present, Athena is implemented mostly in Python, with some parts in Java, and C++. Athena is hosted within the Dynamically-Allocated Virtual Clustering (DAVC) management environment [35], available from our collaborators at the US Army Research Labs. DAVC offers virtual containers that allow easy integration of physical and virtual nodes in the same environment, and thus straightforward migration from emulation to deployment. DAVC currently runs on an Ubuntu server (a 64-bit machine) in emulation mode, and has dependencies on NTP (for time management) and NFS (for file management). The Athena server implements three main data structures, local/remote query logs, fetch/prefetch queues, and the interest table. They are are asynchronously protected and shared among the functional component threads. A node's main event loop simply waits on a TCP socket for incoming messages, and dispatches received messages, according to their headers, to spawn the corresponding functional threads. We simulate a network by running the actual communication protocol stack in a separate process per emulated node. Each emulated network node is thus uniquely identified by its IP:PORT pair. Below, we describe in more detail the functional details of different Athena threads.

7.1. Query Requests

In this implementation, a user can issue query request(s) at any Athena node, using a Query_Init call. At each node, upon user-query initiation, Athena translates the query into the corresponding Boolean expression over predicates, and starts carrying out necessary predicate (label) evaluation. This processing is done in the context of Query_Recv. The component reacts to received queries (either initiated locally or propagated from neighbor nodes) by carrying out the following execution steps: (i) add the new query to the set of queries currently being processed by the node, (ii) determine the set of sources with relevant data objects using a semantic lookup service [36,37], (iii) compute the optimal source subset using a source selection algorithm [38], (iv) send the Boolean expression of the query to neighbors and (v) use a decision-driven scheduler to compute an optimal object retrieval order according to the current set of queries. Requests for those objects that are slated for retrieval are then put in a queue, called the fetch queue. Note how, in this architecture, a node can receive the Boolean expression of a query from step (iii) above before actually receiving requests for retrieving specific objects. This offers an opportunity to prefetch objects not yet requested. A node receiving a query Boolean expression from neighbor nodes will try prefetching data objects for these remote queries, so these objects are ready when requested. Such object requests are put in a prefetch queue. The prefetch queue is only processed in the background. In other words, it is processed only when the fetch queue is empty. When a queue is processed, an object Request_Send function is used to request data objects in the fetch/prefetch queue from the next-hop neighbors.

7.2. Data Object Requests

As a query is decomposed into a set of data object requests, each corresponding to a specific label to be resolved. These requests are then sent through the network towards their data source nodes. Each node maintains an Interest Table that keeps track of which data objects have been requested by which sources for what queries. The interest table helps nodes keep track of upstream requests and avoid passing along unnecessary duplicate data object requests downstream.

Each node also serves as a data cache, storing data objects that pass through, so new requests for a piece of data object that is already cached can be served faster. When a forwarder node already has a cached copy of a piece of data, it needs to decide as to whether or not this cached copy is still fresh enough to serve an incoming request for this piece of data. If yes, then the forwarder would just respond to this request by returning the cached object, otherwise it would pass along the request towards the actual source for a fresh copy.

Specifically, a Request_Recv is called upon receiving an object request from a neighbor. The request is first bookmarked in the interest table. Then, if the object is not available locally, the request is forwarded (using Request_Send) closer to the data source node if the request was a fetch (prefetch requests are not forwarded).

Above, we just discussed how data object requests are handled by Athena nodes. Next, we will look at how Athena handles the transmission of the actual requested data content, either from actual data source nodes or intermediate nodes upon cache hits, back towards the requesters.

7.3. Data Object Replies

Requested data objects (e.g., a picture, an audio clip, etc) are sent back to corresponding requesters in the similar hop-by-hop fashion as that of the requests themselves. Each data object, as it is being passed through intermediate forwarder nodes, is cached along the way. Cached data objects will decay over time, and eventually expire as they reach their freshness deadlines (age out of their validity intervals). In terms of functional interfaces, each Athena node implements the following two functions: Data_Send is used to send requested data object content back towards the original requesters; and Data_Recv is invoked upon receiving a piece of requested data object, which is then matched against all entries in the interest table. If the current node is the original query requester node, the data object is presented to the user for the label value, which is in turn used to update the query. Otherwise, the object will be forwarded to the next hop towards the original requester.

One important note here is that in Athena, a raw data object needs to be sent from the source back to the requester only when the predicate evaluation (labeling) has to be done by the requesting source. For example, after an earthquake, a user is using Athena to look for a safe route to a nearby medical camp. In doing so, Athena retrieves road-side pictures along possible routes for the user to examine. This judgment call—looking at a picture and recognizing it as a safe or unsafe road segment—is put in the hands of the user (the human decision-maker) at the original query requester node. Alternatively, predicate evaluation could be made by machines automatically (e.g., using computer vision techniques to label images). If a qualified evaluator is found at a node for a given predicate, the predicate can be evaluated when the evidence object reaches that node. If the source of the query specified that the signature of this evaluator is acceptable, only the predicate evaluation is propagated the remaining way to the source (as opposed to the evidence object). In the implementation, we restrict predicate evaluators to sources of the query.

7.4. Label Caching

As requested data objects arrive, the query source can then examine the objects and use their own judgment to assign label values to the objects for the particular query task. These labels are injected back into the network, such that future data requests might potentially be served by the semantic labels rather than actual data objects, which depends on whether the requests need to evaluate the same predicates, and what trust relations exist among the different entities (e.g., Alice might choose not to trust Bob's judgment, and thus would insist on getting the actual data object when a matched label from Bob already exists). As such human labels are propagated from the evaluator nodes back into the network towards the data source nodes, they are cached along the way, and can be checked against the interest tables and, upon matches, used locally to update query expressions, and forwarded to the data requesters. Compared to sending actual data objects, sharing and utilizing these labels can lead to several orders of magnitude resource savings for the particular requests.

To help better visualize how the various discussed components work together, we show, in Figure 3, an example of requests and data flows for a particular query.

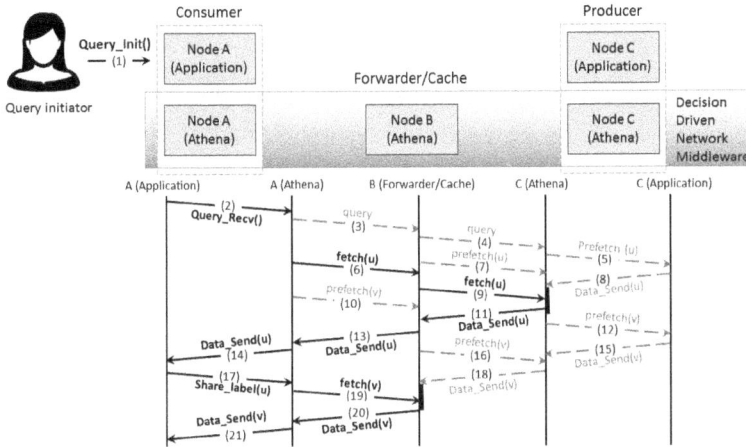

Figure 3. A visualization showing the flow of requests and data as nodes in Athena work together to resolve a query. In this example, the user uses Query_Init() to create and issue a query at Node A. The query in our example involves two data objects, *u* and *v*. Node A calls Query_Recv() locally to start the processing. The query is propagated through the network (Edges 3 and 4), reaching Node B and C. Upon receiving the query, Node C attempts prefetching, first for data object *u* in this particular example. Since Node C is the data source for *u*, it sends *u* back towards the requester node (Edges 8, 11, 13, and 14), during which, Node A's fetch request meets the returned data at Node C (Edge 9). Upon receiving *u* at Node A, the user examines the data, makes a judgment regarding the corresponding condition state of the query. This state label provided by the human decision-maker is then propagated back into the network (Edge 17). The handling for data object *v* follows a similar pattern, for which the fetch request has a cache hit at the forwarder Node B, without reaching the actual source Node C, due to prefetch requests. In the figure, grey arrows and requests represent those processed in the background; namely pre-fetches and their responses. Solid black arrows and requests are those processed in the foreground; namely actual object fetch requests and their responses.

8. Evaluation

In our prototype Athena implementation, a single process—a multithreaded Athena server—was instantiated per emulated node. The server's main event loop simply waited on a TCP socket for incoming messages, and dispatched received messages, according to their headers, spawning the corresponding functional threads as needed. A communication server implemented the desired protocol stack. The application was compiled together with the Athena server. Each emulated network node was uniquely identified by its IP:PORT pair, referring to the server's machine IP and port address associated with the emulated node.

Since we have not had the opportunity to test our system in an actual post-disaster environment, we adopted a set of simulation-based experiments featuring a post-disaster route-finding scenario, where Athena is deployed in a disaster-hit region and is used by people in the region to carry out situation assessment and route-reconnaissance tasks. For simplicity, we consider a Manhattan-like map, where road segments have a grid-like layout. The EMANE-Shim network emulator [39,40] was used to handle all data object transmissions among Athena server processes.

More specifically, we divided the experimental region into a Manhattan grid given by an 8×8 road segment network, with around 30 Athenanodes deployed on these segments, where each node's data can be used to examine the node's immediate surrounding segments. Data objects range from 100 KByte to around 1 MByte, roughly corresponding to what we might expect from pictures taken by roadside cameras. The network emulator is configured with 1 Mbps node-to-node connections. Each route-finding query consists of 5 candidate routes that are computed and randomly selected from the underlying road segment network. Additionally, each node is issued three concurrent queries. With these generic parameter settings, we next discuss each set of the experiments, where the corresponding particular parameter settings will be specified. For collecting results, each data point is produced by repeating the particular randomized experiment 10 times. As resource constraints is our main optimization goal, we use resource consumption (network bandwidth usage) as the main evaluation metric, unless otherwise specified in particular sets of experiments. Table 1 summarizes the key settings.

Table 1. Key emulation parameter settings (unless individual experiments state otherwise).

Network nodes	30
Data object size	100 KB–1 MB
Link bandwidth	1 Mbps node-to-node
Viable options per decision	5
Queries per node per experiment	3
Number of experiments per graph point	10

For the information retrieval schedule, we experimented with multiple baselines, besides our own algorithm, Algorithm 1, introduced in Section 4. All compared algorithms are listed as follows:

- Comprehensive retrieval (cmp): As a first baseline, we include a simple algorithm where all relevant data objects for each query are considered for retrieval. An object is relevant if it provides evidence regarding at least one of the unknowns specified in the decision logic. Note that, comprehensive retrieval does not try to minimize potential redundancy in retrieved evidence, nor does it optimize retrieval order based on considerations such as freshness constraints, deadlines, and cost.
- Selected sources (slt): This is one step beyond the above cmp baseline, where data source selection is performed to minimize redundancy in retrieved evidence. Specifically a coverage problem is solved to obtain the candidate set of data objects to be retrieved that allow evaluating all predicates in the underlying decision logic. Thus, if two cameras have overlapping views of the same road segment, source selection will typically choose only one to retrieve data from. We borrow a state of the art source selection algorithm [38] and use it in our implementation and experiments.
- Lowest cost source first (lcf): This scheme takes the above selected source nodes, and sorts them according to their data object retrieval costs (i.e., data object size), prioritizing objects with lower costs.
- Variational longest validity first (lvf): Our scheduling algorithm, as discussed in Section 4, except that values of labels are not propagated into the network for future reuse.
- Variational longest validity first with label sharing (lvfl): Our scheduling algorithm, with label sharing enabled. Therefore, after a piece of retrieved evidence is annotated with a label, this label information is propagated back into the network towards the corresponding data source node.

Thus, any node along the path that intercepts a future request for this data object can return this label value rather than (requesting and) returning the actual data object.

With these above five different information retrieval scheduling schemes, we carry experiments to study Athena's behavior along the following different dimensions.

- Environment dynamics: We experiment with different levels of environment dynamics, where different portions of data objects are considered to be of fast/slow-changing nature (i.e., having short/long validity intervals).
- Query issuance pattern: We experiment with how queries are issued to nodes–more specifically, whether all queries are issued to only a few nodes or a large number of them.
- Query complexity: For each query, we experiment with varying number of candidate routes, and different number of road segments per route. These correspond to the number of courses of action that are OR'ed together, and the number of predicates AND'ed to establish viability of each course, respectively.
- Query interest distribution: We experiment with two scenarios, namely whether all inquiries are focused on a small hotspot or spanning the entire global region.
- Query locality: We experiment with how "localized" queries are. Basically, a localized query inquires about a node's immediate surrounding area, whereas a more diverse query may ask about data objects on the far end of the network.
- Network topology: We generally use randomly generated network topologies for our experiments. However, we also experiment with two other specific network topologies, namely linear and star shaped, to see how different patterns of network connectivity might affect system behavior.

Data objects range from 100 KByte to around 1 MByte, roughly corresponding to what we might expect from pictures taken by roadside cameras. The network simulator is configured with 1 Mbps node-to-node connections, which is what one might expect when fast wired infrastructure has been destroyed by a disaster, resulting in slower ad hoc links. Each route-finding query consists of five candidate routes that are computed and randomly selected from the underlying road segment network. With these generic parameter settings, we next discuss each set of the experiments, where the corresponding particular parameter settings will be specified. For collecting results, each data point is produced by repeating the particular randomized experiment 10 times.

First and foremost, as our Athena information management system is designed for situation assessment and decision-making under dynamic post-disaster environments, we look at how its query resolution capability is affected by different levels of environment dynamics. In our experiment, data objects generally belong to two different categories, namely slow changing and fast changing. For example, a blockage on a major highway might get cleared within hours, but a damaged bridge likely will take days/weeks or even longer to repair. In this set of experiments, we explore how different mixtures of slow and fast changing objects affect the performance of each of the information retrieval schemes. The results are shown in Figure 4. As seen, at all levels of environment dynamics, our data-validity aware information retrieval schemes are able to successfully resolve most, if not all, queries (i.e., perform them on time and based on fresh data), whereas the baseline methods struggle even with a relatively low level of environment dynamics (an incorrectly resolved query, for our purposes is one where the decision missed the deadline or was based on data that passed their validity interval). This is due to their failure to take into account the deadline and data validity information when scheduling retrievals, which then leads to data expirations and refetches or deadline misses. This not only increases bandwidth consumption, but also prolongs query resolution process, potentially causing more data to expire.

The actual network bandwidth consumption comparisons of all schemes are shown in Figure 5. We already saw from Figure 4 that the various baseline schemes fall short in terms of query resolution ratio; here we observe that they additionally consume more network bandwidth. Comprehensive retrieval scheduling incurs the highest amount of network traffic, as it neither is careful about avoiding redundant data object retrieval, nor tries to follow a meaningful order when fetching data. Network bandwidth consumption marginally decreases as we include source selection (slt) and then follow a lowest-cost-first (lcf) data retrieval schedule. None of the above schemes take into consideration environment dynamics. Therefore, they tend to result in more information expiration and refetching, leading to extraneous bandwidth usage. This additional usage is effectively minimized/avoided by our scheduling strategy, which leads to a considerable decrease in network bandwidth consumption. Additionally, when opportunistic label sharing (lvfl) is enabled in Athena, more significant bandwidth savings are observed, since labels are transmitted instead of actual data objects when possible.

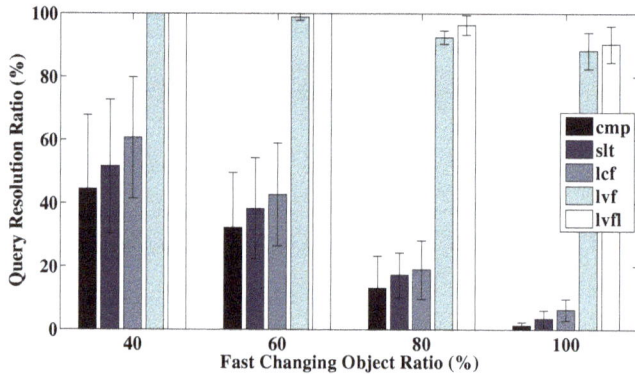

Figure 4. Query resolution ratio at varying levels of environment dynamics (ratio of fast changing objects). cmp, comprehensive retrieval; slt, selected sources; lcf, lowest-cost-first; lvf, longest validity first; lvfl, longest validity first with label sharing.

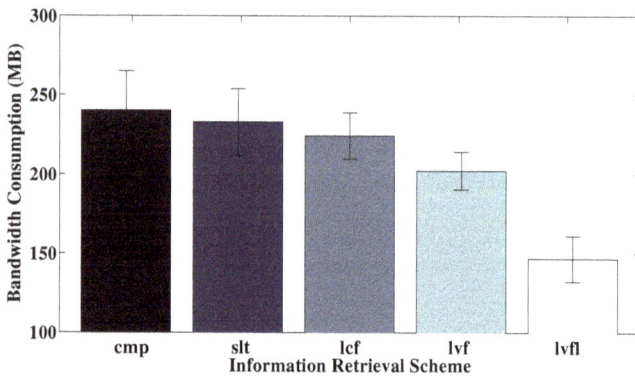

Figure 5. Total network bandwidth consumption comparison of all schedule schemes (with 40% fast changing objects).

Next, we take a look at how query issuance patterns affect system performance. This essentially means how many nodes of the entire network users are issuing queries. If the number is 1 (which is less likely for realistic scenarios), this essentially is equivalent to a centralized application where all users issue all queries at a central node. Under more realistic settings, however, this number will be high as users all around the region would potentially need to request information at each of their own respective locations. We experiment with issuing queries by a single, a pair, half, as well as all nodes. The results are shown in Figure 6. First of all, we observe that the more centralized the query issuance pattern is, the lower the network bandwidth consumption. This makes sense because, given the same number of random queries, having fewer query nodes means higher chance of cache hits for both data objects and shared labels. We also observe that, as query issuance pattern shifts from centralized towards distributed, our information retrieval schemes lead to better improvements over baseline methods as well as consistently stable performance on query resolution ratio.

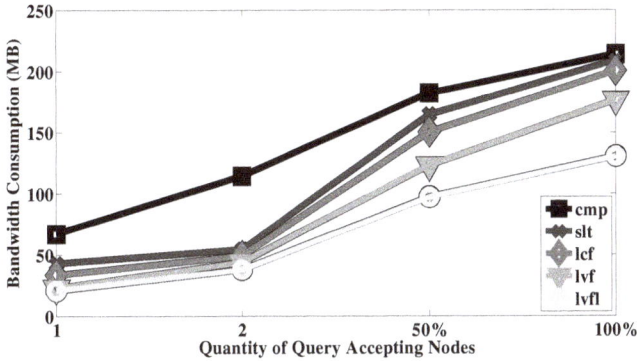

Figure 6. Different query issuance patterns (i.e., varying number of nodes to which a fixed number of queries are issued to.

Next, we look at how query complexity affects system performance. As each query is represented in its logical form, an OR of ANDs, we can vary the number of ANDs under the OR, as well as the number of tests that needs to be performed for each AND in our route finding scenario. This naturally corresponds to the number of candidate routes for each query, and the number of road segments for each candidate route. As shown in Figure 7, increasing the number of routes per query leads to higher network bandwidth consumption for all information retrieval schemes. It is worth noting that, our decision-aware scheduling algorithms (lvf, lvfl) again lead to a slower bandwidth increase compared to other baseline methods, thanks to their ability to exploit queries' internal structure and prune logical evaluations, which would otherwise lead to unnecessary network traffic.

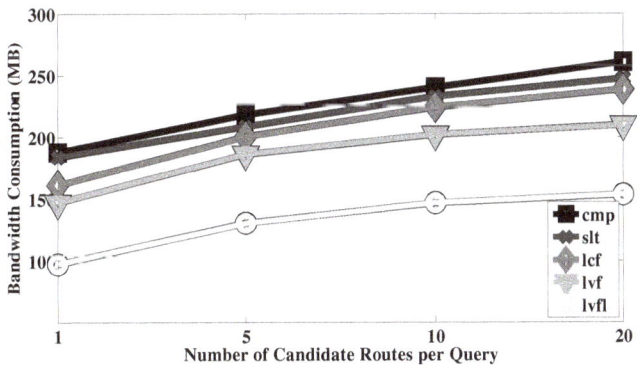

Figure 7. Varying number of candidate routes per query.

Figure 8 shows the experiment results with varying route lengths, where routes are categorized into 5 different length percentiles. As seen, the general relative comparison in terms of network bandwidth consumption remains the same as that of previous experiments. However, we notice an interesting convex shape rather than a monotonic trend (i.e., longer route lengths lead to higher network bandwidth consumption) that some might have expected. The reason lies in the fact that the higher the percentile, the lower the number of route choices there are—i.e., we can easily find in the road network a large number of different short routes, but there might be few options for extremely long ones. This lack of route choices then leads to a higher number of repeated road segments in the queries, which in turn leads to higher cache hit rates and thus fewer transmissions of data objects from their original source nodes. Therefore, this set of experiments also illustrates how queries' interest distributions affect system performance; namely, all other conditions being equal, the more concentrated the queries' interests are, the lower the system bandwidth consumption is. This is due to higher cache hit rates for data objects as well as shared labels values.

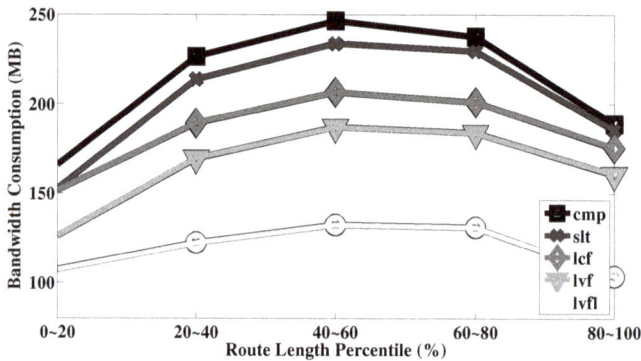

Figure 8. Varying length ranges of queries' intended routes.

Next, we examine how queries' locality can affect system performance. Here, locality refers to how close a query's interests are from the node where it is issued. A nearby query is likely only expressing interests in the vicinity of its issuance node, whereas a faraway query might inquire about data objects located at the far end of the network. Experimental results on network bandwidth consumption are shown in Figure 9. We again omit reiterating the comparison between the different information retrieval scheduling schemes as it is similar to that of previous experiments. We do want to point out the non-monotonicity of bandwidth comparison as we move from nearby to random and then to faraway queries. First of all, when each query is only interested in its close vicinity, the query itself is often of low complexity (i.e., containing fewer candidate routes, and fewer road segments per route), and few data objects need to travel long paths to reach their requesters. Thus, the overall network bandwidth usage remains low. None of these mentioned characteristics still hold when queries' interests shift from nearby to randomly covering the entire region, causing much higher traffic in the network. Then finally, when we further shift queries' interests to be only focusing on faraway objects, we have actually limited the object interest candidate pool, causing queries to overlap more on their relevant data object set. This too leads to higher cache hit rate for both data objects and human labels, similar to what we observed in Figure 8.

Finally, we take a brief look at how different network topologies affect system performance. In addition to randomly generated topologies, we also experiment with linear and star shape networks as two vastly different topologies in terms of network diameter. Results are shown in Figure 10. As seen, networks where nodes are linearly connected result in significantly higher bandwidth consumption. This is understandable because of the excessively long paths data packets need to travel to reach their destinations. The higher probability of data validity expiration caused by long retrieval latencies overshadows the benefits of cache hits along the long paths. The star shape networks, on the other hand, have short transmission paths, but the network bandwidth consumption is not much lower than that of linear networks, due to the existence of a bottleneck at their star centers, which lead to network congestion. This, in turn, causes more data to expire and hence excessive refetches. It is promising to see that in general random network topologies, Athena, with our retrieval schedule schemes, gives good performance.

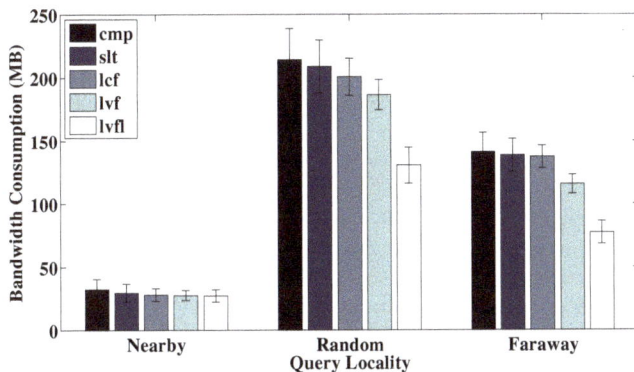

Figure 9. Different levels of queries' localities.

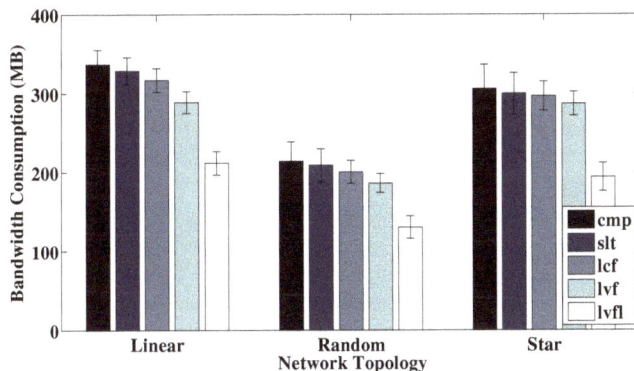

Figure 10. Different network topologies.

9. Discussion

The bulk of this paper focused on challenges in decision-driven execution that are more directly aligned with distributed computing. However, the paradigm offers interesting research opportunities in other related areas, as well. For example, the paradigm offer a mechanism for networks to learn about their users and the physical environment, then exploit such learned knowledge for optimizing decision-making. The decision model used by the network can itself be refined over time as the system observes the decision-makers' information requests, records their decisions and takes note of the underlying context as measured by the multitude of sensors connected to the system. Subsequent mining of such data can lead to progressive refinement of decision-making models and to increasingly accurate reverse-engineering of decision strategies of individuals and groups. Such learned knowledge can, in turn, be applied to optimize the cost of future decision-making. The system, being connected to sensors, can also derive its own models of physical phenomena over time using any of an array of well-known estimation-theoretic approaches. These models can inform settings of various elements of object metadata, such as validity intervals of different types of measurements and probability distributions of particular observed quantities.

While much of the discussion in this paper focused on using the structure of a single decision query to anticipate future object fetch requests, it is also possible to apply pattern mining techniques to identify common decision query sequences and thus anticipate not only current, but also future decision needs. This is possible because users, in many cases, adhere to prescribed workflows dictated by their training, standard operating procedures or doctrine. The workflow is a flowchart of decision points, each conditioned on certain variables or inputs. Since the structure of the flowchart is known, so are the possible sequences of decision points. One can therefore anticipate future decisions given current decision queries. Anticipating future information needs can break traditional delay-throughput constraints: anticipating what information is needed next, as suggested by mission workflow, gives the system more time to acquire it before it is actually used.

Finally, observe that decisions can be conditioned not only on the current state, but also on an anticipated state. For example, a decision on where to intercept a fleeing criminal will depend on predictions of where the criminal goes next. This information may be inferred indirectly from current measurements. Hence, decision-driven execution lends itself nicely to increasing the efficacy of missions involving a significant anticipatory or prediction component, as it offers the mechanisms needed to furnish evidence supporting the different hypotheses or predictions of future actions of agents in the physical environment. The system can therefore empower applications involving intelligent adversaries, such as military operations or national security applications. The design of such applications on top of decision-driven execution systems remains an open research challenge.

10. Conclusions

In this paper, we outlined a novel paradigm for distributed management of communication and scheduling resources, where all resource consumption is driven by information needs of decision-making. The hallmark of the paradigm lies in exporting the logical inference structure of decision-making to the underlying communication resource management layer in order to enable more efficient acquisition of data that simultaneously increase decision timeliness and lower decision cost, while improving decision quality. We implemented these ideas within a framework called Athena. The framework was evaluated from two perspectives; namely, resource (i.e., communication bandwidth) consumption and ability to make valid and timely decisions. It was shown that the framework increases the latter while reducing the former. The approach, therefore, holds promise in jointly meeting timeliness, cost and accuracy goals. The paper constitutes a first step on a lengthy road. Many challenges remain that are delegated to future work. For example, more attention is needed to modeling uncertainty. Often, the exact decisions to be made in the context of a mission are not formulated precisely ahead of time; the unknowns relevant to them might not be clear; and the sources who have information on these unknowns might not be identified. The relevance of individual sources may further depend on the context, which itself might not be precisely known. More attention is also needed in modeling cost. A large object from a sensor with a faster connection might be less costly to retrieve than a smaller object from a sensor with a very slow link. Since connection speed may vary dynamically, cost itself might not be exactly known or subject to change. Furthermore, the decision models themselves can vary. While we assumed a model given by AND-OR trees (of predicates) over relevant unknowns, more involved multi-layer models are possible and should be considered in future work. The confluence of the above challenges in uncertainty estimation, cost metrics, context measurement and adaptation to various runtime dynamics and decision models significantly complicates the formulation of scheduling problems and the attainment of solutions with clean optimality properties. Hence, much work is needed on the algorithmic and theory side to understand various (near-)optimality and impossibility results in this area. At last, the notions of anticipation embedded in the work suggest a fundamental trade-off between robustness and optimality. Basing decisions on more aggressive predictions (or anticipation models) that make stronger assumptions regarding future states can improve results when these assumptions hold, but fail more severely when assumptions are violated. In contrast, less detailed predictions (that make fewer assumptions regarding the future) tend to hedge bets against the unknown, leading to improved robustness with respect to uncertainty. However, they also result in a somewhat impaired average performance. The authors are currently addressing some of the above challenges with a particular focus on exploring the aforementioned tension between robustness and optimality in the context of decision-centric information delivery.

Acknowledgments: Research reported in this paper was sponsored in part by the Army Research Laboratory under Cooperative Agreements W911NF-09-2-0053 and W911NF-17-2-0196, and in part by NSF under Grants CNS 16-18627, CNS 13-02563, CNS 13-45266 and CNS 13-20209. The views and conclusions contained in this document are those of the authors and should not be interpreted as representing the official policies, either expressed or implied, of the Army Research Laboratory, NSF or the U.S. Government. The U.S. Government is authorized to reproduce and distribute reprints for Government purposes notwithstanding any copyright notation herein.

Author Contributions: Jongdeog Lee and Kelvin Marcus designed and developed the current version of Athena, with advice from Tarek Abdelzaher, Amotz Bar-Noy, Ramesh Govindan and Reginald Hobbs; Tanvir Amin significantly contributed to implementation efforts; Shaohan Hu and William Dron developed a previous version of the system; Jung-Eun Kim and Lui Sha contributed real-time decision-driven scheduling extensions; Shuochao Yao and Yiran Zhao worked on various optimizations.

Conflicts of Interest: The authors declare no conflict of interest. The founding sponsors had no role in the design of the study; in the collection, analyses or interpretation of data; in the writing of the manuscript; nor in the decision to publish the results.

References

1. Zanella, A.; Bui, N.; Castellani, A.; Vangelista, L.; Zorzi, M. Internet of Things for Smart Cities. *IEEE Internet Things J.* **2014**, *1*, 22–32.
2. Khatoun, R.; Zeadally, S. Smart cities: Concepts, architectures, research opportunities. Commun. ACM **2016**, *59*, 46–57.
3. Bordel, B.; Alcarria, R.; Robles, T.; Martín, D. Cyber–physical systems: Extending pervasive sensing from control theory to the Internet of Things. *Pervasive Mob. Comput.* **2017**, *40*, 156–184.
4. De, S.; Zhou, Y.; Larizgoitia Abad, I.; Moessner, K. Cyber–Physical–Social Frameworks for Urban Big Data Systems: A Survey. *Appl. Sci.* **2017**, *7*, 1017.
5. Abdelzaher, T.; Amin, M.T.A.; Bar-Noy, A.; Dron, W.; Govindan, R.; Hobbs, R.; Hu, S.; Kim, J.E.; Lee, J.; Marcus, K.; et al. Decision-Driven Execution: A Distributed Resource Management Paradigm for the Age of IoT. In Proceedings of the 2017 IEEE 37th International Conference on Distributed Computing Systems (ICDCS), Atlanta, GA, USA, 5–8 June 2017; pp. 1825–1835.
6. Srivastava, M.; Abdelzaher, T.; Szymanski, B. Human-centric sensing. *Philos. Trans. R. Soc. A* **2012**, *370*, 176–197.
7. Stankovic, J.A. Research directions for the internet of things. *IEEE Internet Things J.* **2014**, *1*, 3–9.
8. Bu, W.; Tang, A.; Han, J. An analysis of decision-centric architectural design approaches. In Proceedings of the 2009 ICSE Workshop on Sharing and Reusing Architectural Knowledge, Vancouver, BC, Canada, 16 May 2009; pp. 33–40.
9. Cui, X.; Sun, Y.; Mei, H. Towards Automated Solution Synthesis and Rationale Capture in Decision-Centric Architecture Design. In Proceedings of the Seventh Working IEEE/IFIP Conference on Software Architecture (WICSA 2008), Vancouver, BC, Canada, 18–21 February 2008; pp. 221–230.
10. Van Heesch, U.; Eloranta, V.P.; Avgeriou, P.; Koskimies, K.; Harrison, N. Decision-Centric Architecture Reviews. *IEEE Softw.* **2014**, *31*, 69–76.
11. Panchal, J.H.; Gero Fernández, M.; Paredis, C.J.; Allen, J.K.; Mistree, F. A modular decision-centric approach for reusable design processes. *Concurr. Eng.* **2009**, *17*, 5–19.
12. Paschke, A. A semantic rule and event driven approach for agile decision-centric business process management. In *Towards a Service-Based Internet*; Springer: New York, NY, USA, 2011; pp. 254–267.
13. Feng, X.; Richards, G.; Raheemi, B. The road to decision-centric business intelligence. In Proceedings of the IEEE International Conference on Business Intelligence and Financial Engineering (BIFE'09), Beijing, China, 24–26 July 2009, pp. 514–518.
14. Blenko, M.W.; Mankins, M.C.; Rogers, P. The decision-driven organization. *Harv. Bus. Rev.* **2010**, *88*, 54–62.

15. Pecharich, J.; Stathatos, S.; Wright, B.; Viswanathan, A.; Tan, K. Mission-Centric Cyber Security Assessment of Critical Systems. In *AIAA SPACE 2016*; Aerospace Research Central: Reston, VA, USA, 2016; p. 5603.

16. Albanese, M.; Jajodia, S.; Jhawar, R.; Piuri, V. Securing Mission-Centric Operations in the Cloud. In *Secure Cloud Computing*; Jajodia, S., Kant, K., Samarati, P., Singhal, A., Swarup, V., Wang, C., Eds.; Springer: New York, NY, USA, 2014; pp. 239–259.

17. Jajodia, S.; Noel, S.; Kalapa, P.; Albanese, M.; Williams, J. Cauldron mission-centric cyber situational awareness with defense in depth. In Proceedings of the Military Communications Conference (MILCOM 2011), Baltimore, MD, USA, 7–10 November 2011, pp. 1339–1344.

18. Decision-Driven Resource Management Patent. Available online: http://www.google.com/patents/US20150142726 (accessed on 10 November 2017).

19. Liu, S.; Zaraté, P. Knowledge Based Decision Support Systems: A Survey on Technologies and Application Domains. In *Group Decision and Negotiation. A Process-Oriented View, Proceedings of the Joint INFORMS-GDN and EWG-DSS International Conference, GDN 2014, Toulouse, France, 10–13 June 2014*; Zaraté, P., Kersten, G.E., Hernández, J.E., Eds.; Springer: Cham, Switzerland, 2014; pp. 62–72.

20. Vadlamudi, S.G.; Chakraborti, T.; Zhang, Y.; Kambhampati, S. Proactive Decision Support using Automated Planning. *arXiv* **2016**, arXiv:1606.07841.

21. Seligman, L.; Lehner, P.; Smith, K.; Elsaesser, C.; Mattox, D. Decision-centric information monitoring. *J. Intell. Inf. Syst.* **2000**, *14*, 29–50.

22. Pourshahid, A.; Richards, G.; Amyot, D. Toward a Goal-Oriented, Business Intelligence Decision-Making Framework. In *E-Technologies: Transformation in a Connected World: 5th International Conference, MCETECH 2011, Les Diablerets, Switzerland, 23–26 January 2011*; Revised Selected Papers; Babin, G., Stanoevska-Slabeva, K., Kropf, P., Eds.; Springer: Berlin/Heidelberg, Germany, 2011; pp. 100–115.

23. Kim, J.E.; Abdelzaher, T.; Sha, L.; Bar-Noy, A.; Hobbs, R. Sporadic Decision-centric Data Scheduling with Normally-Off Sensors. In Proceedings of the IEEE International Real-Time Systems Symposium (RTSS), Porto, Portugal, 29 November–2 December 2016.

24. Kim, J.E.; Abdelzaher, T.; Sha, L.; Bar-Noy, A.; Hobbs, R.; Dron, W. On Maximizing Quality of Information for the Internet of Things: A Real-time Scheduling Perspective (Invited). In Proceedings of the IEEE International Conference on Embedded and Real-Time Computing Systems and Applications, Daegu, Korea, 17–19 August 2016.

25. Hu, S.; Yao, S.; Jin, H.; Zhao, Y.; Hu, Y.; Liu, X.; Naghibolhosseini, N.; Li, S.; Kapoor, A.; Dron, W.; et al. Data Acquisition for Real-Time Decision-Making under Freshness Constraints. In Proceedings of the IEEE International Real-Time Systems Symposium (RTSS), San Antonio, TX, USA, 1–4 December 2015.

26. Adelberg, B.; Garcia-Molina, H.; Kao, B. Applying Update Streams in a Soft Real-time Database System. In Proceedings of the ACM SIGMOD International Conference on Management of Data, San Jose, CA, USA, 22–25 May 1995.

27. Kang, K.D.; Son, S.; Stankovic, J.; Abdelzaher, T. A QoS-sensitive approach for timeliness and freshness guarantees in real-time databases. In Proceedings of the Euromicro Conference on Real-Time Systems (ECRTS), Vienna, Austria, 19–21 June 2002.

28. Kang, K.D.; Son, S.H.; Stankovic, J.A. Managing Deadline Miss Ratio and Sensor Data Freshness in Real-Time Databases. *IEEE Trans. Knowl. Data Eng.* **2004**, *16*, 1200–1216.

29. Xiong, M.; Wang, Q.; Ramamritham, K. On earliest deadline first scheduling for temporal consistency maintenance. *Real-Time Syst.* **2008**, *40*, 208–237.

30. Tufekci, S.; Wallace, W.A. The Emerging Area Of Emergency Management And Engineering. *IEEE Trans. Eng. Manag.* **1998**, *45*, 103–105.

31. Hu, S.; Li, S.; Yao, S.; Su, L.; Govindan, R.; Hobbs, R.; Abdelzaher, T. On Exploiting Logical Dependencies for Minimizing Additive Cost Metrics in Resource-Limited Crowdsensing. In Proceedings of the IEEE International Conference on Distributed Computing in Sensor Networks (DCoSS), Fortaleza, Brazil, 10–12 June 2015.

32. Wang, S.; Abdelzaher, T.; Gajendran, S.; Herga, A.; Kulkarni, S.; Li, S.; Liu, H.; Suresh, C.; Sreenath, A.; Wang, H.; et al. The Information Funnel: Exploiting Named Data for Information-Maximizing Data Collection. In Proceedings of the 2014 IEEE International Conference on Distributed Computing in Sensor Systems (DCOSS '14), Marina Del Rey, CA, USA, 26–28 May 2014; IEEE Computer Society: Washington, DC, USA, 2014; pp. 92–100.

33. Lee, J.; Kapoor, A.; Amin, M.T.A.; Wang, Z.; Zhang, Z.; Goyal, R.; Abdelzaher, T. InfoMax: An Information Maximizing Transport Layer Protocol for Named Data Networks. In Proceedings of the 2015 24th International Conference on Computer Communication and Networks (ICCCN), Las Vegas, NV, USA, 3–6 August 2015; pp. 1–10.

34. Zhang, L.; Estrin, D.; Burke, J.; Jacobson, V.; Thornton, J.D.; Zhang, S.B.; Dmitri, G.T.K.C.; Massey, K.D.; Papadopoulos, C.; Lan, T.A.; et al. *Named Data Networking (NDN) d NDN-0001*; NSF FIA Kickoff Meeting: Arlington, VA, USA, 2010.

35. Marcus, K.; Cannata, J. Dynamically allocated virtual clustering management system. In Proceedings of the SPIE Proceedings on Ground/Air Multisensor Interoperability, Integration, and Networking for Persistent ISR IV, Baltimore, MD, USA, 22 May 2013; Volume 8742.

36. Ra, M.R.; Liu, B.; La Porta, T.F.; Govindan, R. Medusa: A programming framework for crowd-sensing applications. In Proceedings of the ACM International Conference on Mobile Systems, Applications, and Services (MobiSys), Low Wood Bay, Lake District, UK, 25–29 June 2012.

37. Jiang, Y.; Xu, X.; Terlecky, P.; Abdelzaher, T.; Bar-Noy, A.; Govindan, R. MediaScope: Selective On-demand Media Retrieval from Mobile Devices. In Proceedings of the ACM/IEEE International Conference on Information Processing in Sensor Networks (IPSN), Philadelphia, PA, USA, 8–11 April 2013.

38. Bar-Noy, A.; Johnson, M.P.; Naghibolhosseini, N.; Rawitz, D.; Shamoun, S. The Price of Incorrectly Aggregating Coverage Values in Sensor Selection. In Proceedings of the IEEE International Conference on Distributed Computing in Sensor Networks (DCoSS), Fortaleza, Brazil, 10–12 June 2015.

39. US Naval Research Lab Networks and Communication Systems Branch. *Extendable Mobile Ad-hoc Network Emulator (EMANE)*; U.S. Naval Research Lab: SW Washington, WA, USA, 2016.

40. Dron, W.; Leung, A.; Hancock, J.; Aguirre, M.; Thapa.; Walsh, R. *CORE Shim Design Document*; NS-CTA Technical Report; Network Science CTA: Cambridge, MA, USA, 2014.

MDPI
St. Alban-Anlage 66
4052 Basel
Switzerland
Tel. +41 61 683 77 34
Fax +41 61 302 89 18
www.mdpi.com

Journal of Sensor and Actuator Networks Editorial Office
E-mail: jsan@mdpi.com
www.mdpi.com/journal/jsan

www.ingramcontent.com/pod-product-compliance
Lightning Source LLC
Chambersburg PA
CBHW051851210326
41597CB00033B/5860